BARRON'S

Painless
Biology

Cynthia Pfirrmann
B.S, M.S.

Published by Kaplan, Inc., d/b/a Barron's Educational Series
1515 W Cypress Creek Road
Fort Lauderdale, FL 33309
www.barronseduc.com

ISBN: 978-1-5062-8013-4

10 9 8 7 6 5 4 3 2 1

Kaplan, Inc., d/b/a Barron's Educational Series print books are available at special
quantity discounts to use for sales promotions, employee premiums, or educational
purposes. For more information or to purchase books, please call the Simon &
Schuster special sales department at 866-506-1949.

Contents

How to Use This Book. viii

Chapter One: Science and Biology 1

What Is Science? 1

What Is Biology?. 2

How Do We Explore Biology? 5

Vocabulary. 11

BRAIN TICKLERS—THE ANSWERS 14

Chapter Two: Cells. 15

What Are Cells? 15

Levels of Classification 17

Diversity of Cells. 19

What Are the Parts of Cells?. 22

What Happens When Cells Don't Work Correctly?. 27

Vocabulary. 29

BRAIN TICKLERS—THE ANSWERS 32

Chapter Three: Biological Chemistry 33

What Are Atoms? 33

What Are Molecules?. 35

What Are Chemical Bonds? 36

Water and pH 37

Biological Macromolecules. 39

Specialized Biological Molecules 45

Adenosine Triphosphate (ATP). 46

Vocabulary. 48

BRAIN TICKLERS—THE ANSWERS 51

Chapter Four: Cellular Transport 53

What Is Cellular Transport?. 53

Cell Membrane Structure 53

What Is Passive Transport?. 57

What Is Active Transport? 61

What Happens When Cell Membranes Don't Work Correctly? 63

Vocabulary. 64

BRAIN TICKLERS—THE ANSWERS 67

Chapter Five: Metabolism and Energy 69

The Relationship Between Metabolism and Energy. 69

Energy 70

Photosynthesis 72

Chemosynthesis. 75

Cellular Respiration 76

Energy-storing Macromolecules 80

Vocabulary. 82

BRAIN TICKLERS—THE ANSWERS 85

Chapter Six: Human Anatomy. 87

The Human Body 87

Basic Anatomy and Physiology 87

Planes and Directional Terms. 88

Body Cavities and Membranes 89

Microanatomy and Macroanatomy 90

Skeletal System 91

Muscular System 95

Integumentary System 98

Cardiovascular System 101

Respiratory System 105

Digestive System 108

Urinary System 110

Nervous System 113

Endocrine System 116

Lymphatic and Immune Systems. 119

Reproductive System. 122

Vocabulary. 124

BRAIN TICKLERS—THE ANSWERS 127

Chapter Seven: Cellular Reproduction 129

Asexual and Sexual Reproduction 129

The Cell Cycle 130

Mitosis 131

Meiosis 133

Telomeres 136

Vocabulary 139

BRAIN TICKLERS—THE ANSWERS 141

Chapter Eight: Basic Genetics 143

What Is Genetics? 143

What Are Traits and How Do They Work? 144

What Is DNA and How Does It Work? 146

What Are Chromosomes and How Do They Work? 147

What Are Genes and How Do They Work? 148

What Are Alleles and How Do They Work? 149

Vocabulary 150

BRAIN TICKLERS—THE ANSWERS 152

Chapter Nine: Mendelian Genetics 153

Basic Mendelian Inheritance 153

Gregor Mendel 153

Simple Mendelian Inheritance 155

Punnett Squares 158

Complex Patterns of Inheritance 161

Vocabulary 164

BRAIN TICKLERS—THE ANSWERS 166

Chapter Ten: Molecular Genetics 167

A Brief History of the Study of Genetics 167

Nucleic Acid 169

DNA Replication 172

Vocabulary 179

BRAIN TICKLERS—THE ANSWERS 181

Chapter Eleven: Genetic Disorders 183

    Basics of Genetic Disorders 183

    Dominant Disorders 184

    Recessive Disorders 187

    Sex-linked Disorders 191

    Polygenic and Multifactorial Disorders 192

    Chromosomal Abnormalities 194

    Genetic Screening 195

    Vocabulary. 197

    BRAIN TICKLERS—THE ANSWERS 200

Chapter Twelve: Genetic Technology 201

    Polymerase Chain Reaction (PCR). 201

    DNA Microarrays 202

    Home DNA Tests 202

    mRNA Vaccines 204

    Genetic Engineering 205

    Cloning. 208

    Vocabulary. 210

    BRAIN TICKLERS—THE ANSWERS 211

Chapter Thirteen: Ecological Relationships 213

    Relationships Between Species 213

    Autotrophs and Heterotrophs 214

    Habitat and Niche 215

    Feeding Relationships. 216

    Food Webs 217

    Species Interactions 219

    Symbiotic Relationships 221

    Vocabulary. 222

    BRAIN TICKLERS—THE ANSWERS 224

Chapter Fourteen: Ecology and Planet Earth 225

    Levels of Organization. 225

    Biomes. 228

Communities and Succession 232

Ecological Relationships and Energy Flow .235

Nutrient Cycling in Ecosystems 237

Vocabulary. 241

BRAIN TICKLERS—THE ANSWERS 243

Chapter Fifteen: Population Biology and Human Ecology 245

How Populations Grow and Change. 245

Limits to Growth 247

Human Population Growth 250

Vocabulary. 253

BRAIN TICKLERS—THE ANSWERS 255

Chapter Sixteen: Evolution and Natural Selection 257

Evidence of Life on Earth 259

Charles Darwin 262

Theory of Natural Selection 264

Evidence of Evolution 265

Human Evolution 269

Vocabulary. 272

BRAIN TICKLERS—THE ANSWERS 274

Chapter Seventeen: Human Impacts on Earth 275

The Anthropocene Era 275

Conservation Biology Impacts 276

Human Intellect Impacts 282

Vocabulary. 284

BRAIN TICKLERS—THE ANSWERS 287

Index 289

How to Use This Book

Painless biology? Yes, that's right. Learning about and understanding biology does not have to be the intimidating process that so many students expect it to be. Think about it: biology is the study of life—isn't that what you've been doing for most of your life? I've been teaching high school biology for 25 years, and so many students arrive in my classroom afraid of the subject and sure that biology is beyond comprehension. It is not. Think about what you already know about how life works. You're off to a good start, right? There is definitely science to explore and there is a language that belongs to biology, but always keep in mind that when you do the work to explore, understand, and define life, you are working to increase your understanding of yourself and of the world around you. It's totally worth it! Let's go...

Painless Icons and Features

Painless Biology incorporates the following icons and features to help make learning biology easier.

Interesting documented biological facts about the topics are covered in each chapter. Each Painless Fact helps to illuminate and personalize the information presented in that chapter.

You will find a variety of helpful hints and suggestions that are specific to understanding biology throughout the chapters of this book

Throughout these chapters you will find study skill strategies specific to learning about and exploring biology and the sciences.

When the exploration of a topic is sequential or would benefit from a list of steps, the steps are provided under this feature.

REMINDER

Many topics in this book are presented and then revisited and elaborated upon, or explored in a different context, in later chapters. It may help to go back and review the original information as you begin to apply it to a new biology concept.

CAUTION—Major Mistake Territory!

These warnings may help you avoid common mistakes and misunderstandings. Make sure to read them carefully.

BRAIN TICKLERS

You will find Brain Ticklers throughout and at the end of each chapter of *Painless Biology*. These short quizzes are designed to make sure that you understand and recall the information you just learned. Complete the Brain Ticklers and check your answers using the answer key at the end of the chapter. If you get any wrong, you know that that's a topic that needs to be reviewed.

SUPER BRAIN TICKLERS

Super Brain Ticklers appear at the end of each chapter. These are comprehensive quizzes on all of the topics covered in the chapter. Solve these questions to make sure that you understand all of the chapter topics before moving forward.

Chapter Breakdown

Chapter One is an introduction to science and biology. Given that biology is the "study of life," this chapter defines the characteristics shared by all living things. It also discusses one of the most important instruments used to study biology—the microscope.

Chapter Two explores the study of cells. In this chapter you will investigate how organisms are categorized and named. You will explore some of the diversity of life forms. You will examine the structures, called *organelles*, that make up all living things. You will also discover some of the impacts of cells and organelles that do not work correctly.

Chapter Three is about biological chemistry and explores atoms, molecules, and chemical bonding. The importance of water, along with the associated role of pH in living things, is discussed. This chapter explains the biological macromolecules: carbohydrates, lipids, proteins, and nucleic acids. It also explores some important specialized biological molecules like ATP and enzymes.

Chapter Four explores how materials are transported into and out of cells. Cell membrane structure is discussed as well as its impact on transport. Passive and active transport are reviewed including osmosis and osmotic solutions. Finally, this chapter discusses what happens when cell membranes don't work correctly.

Chapter Five looks into the relationship between metabolism and energy. Photosynthesis, chemosynthesis, and cellular respiration and their cycles are explained. Energy-storing macromolecules like ATP are discussed.

Chapter Six is about human anatomy and includes explanations of basic anatomy and physiology, planes and directional terms, body cavities and membranes, and microanatomy and macroanatomy. This chapter explores the basic functions, microscopic and gross anatomy, and diseases of each of the human body systems. All body systems are reviewed: skeletal, muscular, integumentary, cardiovascular, respiratory, digestive, urinary, nervous, endocrine, immune, lymphatic, and reproductive.

Chapter Seven explores cellular reproduction, focusing on the roles of asexual and sexual reproduction. The cell cycle is explained with a focus on mitosis and meiosis. Aging and cellular life spans are discussed in the context of the activity of telomeres.

Chapter Eight is a chapter on the basics of genetics. Here, you will explore what genetics is, what it does, and how it impacts each of us. This chapter explores traits, DNA, chromosomes, genes, and alleles.

Chapter Nine examines basic Mendelian inheritance and its patterns. You will explore a bit of the life and works of the "Father of Modern Genetics," Gregor Mendel. You will explore simple

Mendelian inheritance patterns: recognizing dominant and recessive genes as well as homozygous and heterozygous inheritance. You will learn about and practice with basic Punnett squares. You will also learn about non-Mendelian, or complex, patterns of inheritance.

Chapter Ten is about molecular genetics. You will explore the history of the study of genetics. This chapter will explain nucleic acids, DNA, and RNA. You will learn about how DNA replication, transcription, and translation work.

Chapter Eleven explains and describes the basics of genetic disorders, including dominant, recessive, and sex-linked genetic disorders. You will explore polygenic and multifactorial genetic disorders. The chapter will explain chromosomal abnormalities, and why they are not the same as genetic disorders. Genetic and chromosomal disorder screening tests and their uses will also be explained.

Chapter Twelve is about genetic technology. In this chapter you will learn about PCR, DNA microarrays, home DNA tests, mRNA vaccines, genetic engineering, GMOs, gene editing, CRISPR, and cloning.

Chapter Thirteen will explain the basic biology of ecological relationships. It will explore relationships between species, the importance of habitat and niche, and the roles of autotrophs and heterotrophs. This chapter will also describe feeding relationships, food webs, species interactions, and the different types of symbiotic interactions.

Chapter Fourteen is about ecology and the planet Earth. You will look into ecological levels of organization, the roles and importance of the Earth's biomes, and the relationships between communities and ecological successions. You will explore the relationship between ecological relationships and energy flow as well as nutrient cycling in ecosystems.

Chapter Fifteen will move you into an understanding of human ecology and population biology. You will explore the basics of the unique role of human populations in the Earth's ecosystems. The chapter will explain how human populations grow and change, limits to human population growth, and current information on the impacts of human population growth.

Chapter Sixteen is about evolution and natural selection. This chapter will discuss early evidence of life on Earth as well as the life and work of Charles Darwin and his theory of natural selection. You will explore multiple forms of evidence for evolution and learn about the history of human evolution.

Chapter Seventeen discusses and describes the unique impacts of the human race on the Earth. You will learn why some scientists are labeling this time on Earth the "Anthropocene Era." You will explore the potential positive impacts provided by the field of conservation biology and how human intellect and choices have the potential to reduce negative human-generated destruction of the Earth's resources.

And so here we go…. *Painless Biology* is ready to help you move forward in your understanding of life: how it began and continues on, how living things are organized and assembled, how living things have adapted and changed, how living things obtain and use energy, how living things interact with their living and nonliving surroundings, and how life is explored and understood. Let's learn about biology.

Chapter 1

Science and Biology

What Is Science?
Science is a way of thinking

Have you ever heard people say that they believe something is true or false? Have you thought, "What does that mean?" Did you wonder, "Where did they get their information?" Did you think about asking, "What kind of evidence supports this belief?" Well, if you asked those questions, you were thinking scientifically. Science is an approach to life that involves gathering tested and supported information in order to answer questions and solve problems.

Scientific thinking: the process of investigation

So, how do we go about gathering information that is tested and supported? The first step is to ask questions that are based on observable evidence. You've probably been doing this for years! Kids start asking questions like "Why is the sky blue?" and "Why is the grass green?" when they're about three years old. These are reasonable scientific questions, but keep in mind that questions like "What kind of phone should I buy?" or "What classes should I take next semester?" or "Where should I go to college?" can also be answered effectively using a scientific strategy.

Scientific thinking: the process of evaluating information

Whatever your questions may be, however many types of evidence you collect, and however much evidence you gather, you then have to evaluate the credibility of the evidence. Many people will use the phrase

"I have a theory about...," and in scientific terms this is inaccurate. Usually, it means a person has a guess to make about the topic. In science, a **theory** is a very thoroughly tested and supported explanation of a subject. A scientific theory has been observed and experimented upon many times and has been repeatedly confirmed. A more scientifically accurate term would be **hypothesis**. When someone has a hypothesis, it means that he or she has a suggested answer to a question based on observation and limited information. A hypothesis is a starting point for his or her exploration of the topic. Scientific experimentation, or testing, can only support or refute a hypothesis; it cannot prove or disprove a hypothesis. The word **law** is also used differently in science than in everyday conversation. In science, a law is a statement of fact, established as fact because a specific natural event always occurs given the same conditions.

CAUTION—Major Mistake Territory!

Do not use the word *theory* to mean "I have an idea about that." In scientific conversations, discussing a theory indicates that you are talking about a concept that has been extensively examined, tested, and supported.

Scientific thinking: why does it matter?

In seeking the answers to questions, scientific thinking and scientific processes are used to recognize and identify the importance of thoughtful exploration of these questions. Scientific thinking requires the use of reliably tested and supported data when making scientific conclusions and important decisions. Collectively, these processes allow members of society to make responsible and informed decisions and choices that are supported by reliable evidence.

What Is Biology?

Biology is the science of life

There are many subdivisions of science. There is chemistry, which explores the structure and function of matter. There is physics, which studies how matter works within the universe. Biology studies how life and living things work. Biology explores the chemistry and molecular and microscopic structures of living things. It studies the macroscopic, or visible, structures and functions of living things and how they work together. Biology explores the development, evolution, and ecological interactions of all living things. In short, biology is the complete study of all living things.

Biology studies types of life

Often, when people refer to living things, they are talking about animals and plants. However, the groups of living things actually include bacteria, protists, and fungi as well as animals and plants. Human beings are members of the animal kingdom.

Biology recognizes the characteristics of life

All—yes, ALL—living things share several basic characteristics. Every species will have many other unique characteristics, but these specific characteristics are shared by every single living thing. If something (like water, fire, or a virus) does not have all these characteristics, it is not considered to be living.

1. **Growth**—All living things gain size (length, height, circumference, weight, mass) over time. Growth is not the same as development.

2. **Development**—All living things change as they age (metamorphosis of a tadpole into a frog; puberty and loss of baby teeth in mammals; plants growing seeds and flowers).

3. **Organization**—All living things have cell and body parts performing specific jobs in specific locations (cell membranes always contain cytoplasm in bacteria; the human heart and lungs are always in the chest cavity).

4. **Reproduction**—All living things have the ability to reproduce. Asexual reproduction produces identical new offspring for some species and identical new body cells for multicellular organisms. Sexual reproduction creates new and unique offspring of different parents.

5. **Homeostasis**—All living things have the ability to keep their bodies in balance. Examples of homeostasis include plants

growing toward light, and birds and mammals maintaining their body temperature, pulse, and respiration.

6. **Cells**—All living things either are cells or are made of cells. All cells eventually die.

7. **Metabolism**—All living things obtain nutrients, then they use those nutrients for energy and structural materials, and finally they dispose of the leftover waste.

8. **Adaptation**—All living things adjust to their environment. This does not change their DNA. However, as members of a population, individuals with the adaptations that best meet the demands of their environment will have the most reproductive success. Their better adapted offspring will then contribute to changes in the general population in a process known as evolution.

 BRAIN TICKLERS   Set # 1

Decide whether each of the following statements is true or false.

1. A theory and a hypothesis are each a type of guess.

2. All living things share the characteristic of movement.

3. Growth and development are different characteristics of life.

(Answers are on page 14.)

 PAINLESS TIP

Although there is some controversy about it, most biologists agree that viruses are not living things. Viruses can't create energy for themselves, don't maintain homeostasis, don't grow or develop, and aren't made of cells. Although viruses do meet several criteria for living things, like change due to natural selection and reproduction, they don't have all the characteristics of life. They reproduce by taking over the working parts of living cells. Therefore, they can't be considered living things. Viruses are actually more like parasitic robots than living organisms.

How Do We Explore Biology?

We use the scientific method

Scientists have been using a common model for exploring, examining, and sharing information about the natural world for thousands of years. This model is now called the **scientific method**. At one point, the scientific method was promoted as a linear process that allowed scientists to effectively gather and share scientific information.

PAINLESS STEPS

Identifying the steps that make up the scientific method is painless and looks like this:

Step 1: Observations—Experience the world around you by using all your senses.

Step 2: Questions—Identify the things you wonder about and develop your questions.

Step 3: Hypothesis—Develop a testable and falsifiable statement about your question.

Step 4: Research—Explore what has already been understood about your hypothesis.

Step 5: Experimentation—Develop and perform experiments to test your hypothesis.

Step 6: Data collection—Gather data from the experiments you perform.

Step 7: Analysis—Inspect and examine experimental results.

Step 8: Conclusion—Explain experimental data and their relationship to the original hypothesis. Data will only support or refute the hypothesis; it will not "prove" or "disprove" anything in the hypothesis.

Today, scientists use all these elements of the scientific method, but they understand that their sequence is very flexible. For example, if scientists gather data that do not support their hypothesis, they don't need to start over. Instead, they are likely to modify their questions, hypotheses, research, or experiments to be more appropriate to their investigation.

PAINLESS TIP

A great way to start thinking scientifically is to pay attention to your observations. Then you can start formulating questions about your observations. For example, if you go to a park, or even your yard, and stay still, you may observe the birds interacting. What do they look like? Do they seem to notice each other? What are they doing? Do they seem to have a hierarchy (are some acting dominant)? Do birds that look similar act differently toward each other than birds that look different? How might you begin to look for answers to your questions?

We identify types of information: observation vs. inference

Scientists have to be careful about introducing their own **bias**, or prejudice, into their investigations. They need to objectively evaluate their thoughts and work to make sure they're not introducing false information based on personal bias. One way to do this is to carefully determine whether a statement is based on **observation** or **inference**. Observation is the act of monitoring or carefully examining something in order to get information. Observation is used in scientific reasoning because it stands alone and isn't influenced by personal beliefs. Inference is the use of past experiences, assumptions, or speculation to frame a belief. Scientists avoid the use of inference because it may insert inaccurate or untrue information into their work.

We identify types of data: quantitative vs. qualitative evidence

When we're investigating a scientific question and gathering information, that information is called **data**. Data are generally accepted as being either **quantitative** or **qualitative evidence**. Quantitative data are numerical—for example, measurements of time, temperature, size, mass, and, obviously numbers. Qualitative data are sensory information. Qualitative data include colors, textures, shapes, smells, and tastes. Identifying what kind of information they're looking for helps scientists watch for it more carefully.

BRAIN TICKLERS   Set # 2

Name the term that represents each of the following descriptions.

1. Introducing one's own perspective, or prejudices, into a scientific investigation

2. Evidence that is gathered using the senses (sight, sound, touch, smell, taste)

3. The process of carefully examining something

A. qualitative

B. observation

C. bias

(Answers are on page 14.)

We communicate our data: tables, graphs, and variables

As scientists gather data, they need to arrange and present it in an organized way. Tables and graphs help with this strategy. A **table** is a collection of information arranged into groups or columns. Here is an example of a table:

Vancouver, British Columbia: When to See Wildlife

	A	Bears	Dolphins	Whales
1		Bears	Dolphins	Whales
2	Jan	8	150	80
3	Feb	54	77	54
4	Mar	93	32	100
5	Apr	116	11	76
6	May	137	6	93
7	Jun	184	1	72

Figure 1–1. Example of a Table

A **graph** is a diagram that shows the relationship between two, often quantitative, variables. Line graphs are often used to convey scientific data. A **variable** is a factor in an experiment or investigation that is likely to change. There are different types of variables in an experiment. **Control variables** show us what happens if nothing is changed;

scientists usually measure their results against a control. An **independent variable** is a single factor that changes in an experiment. It can be changed by a scientist or simply by nature. A **dependent variable** is a change that occurs because of the changed independent variable.

PAINLESS TIP

Remember that dependent variables are dependent upon changes in the independent (or manipulated) variable. Dependent variables are usually the results we look for in an experiment.

The line graph below shows a population count over the course of 125 years. The variables here are time (measured in years) and the population count (measured in millions of individuals). Graphs are always titled and the axes are always labeled. In this graph the *x*-axis (horizontal line) is labeled with the years (quantitative data and independent variable) and the *y*-axis (longitudinal line) is labeled with population numbers (quantitative data and dependent variable). This graph tells us that the population change was dependent on the passage of time.

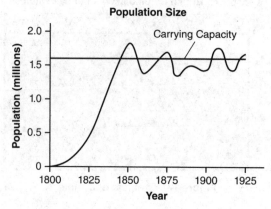

Figure 1–2. Example of a Line Graph

PAINLESS TIP

The most commonly used graphs in biology are based on line graphs. Remember that the line on the bottom, the *x*-axis, is where the independent variable goes. The results, or dependent variable, are graphed on the *y*-axis on the side.

We communicate our data: clarity and measurement

Scientists depend on each other to share information. Scientists also need other scientists to duplicate their experiments and results in order to support or refute their data. For these reasons, it is really important that scientists communicate accurately and clearly. Precise vocabulary matters in science, and so do precise measurements. All scientists around the world use the metric system to gather and share numerical data. The metric system allows these scientists to perform the same experiments and compare data using the same numerical system.

Prefix	In words	Multiply by	Factor
nano (n)	Billionth	1/1,000,000,000	1×10^{-9}
micro (μ)	Millionth	1/1,000,000	1×10^{-6}
milli (m)	Thousandth	1/1,000	1×10^{-3}
centi (c)	Hundredth	1/100	1×10^{-2}
deci (d)	Tenth	1/10	1×10^{-1}
		1	
deca (da)	Ten	10	1×10^{1}
hecto (h)	Hundred	100	1×10^{2}
kilo (k)	Thousand	1,000	1×10^{3}
mega (M)	Million	1,000,000	1×10^{6}
giga (G)	Billion	1,000,000,000	1×10^{9}

Figure 1–3. Metric Conversion Chart

We use technology to investigate: microscopy

One type of technology that biologists are known to use is the microscope. Microscopes allow us to see objects that are too small to see with just our eyes. Microscopes range from very simple to incredibly intricate in terms of complexity. The type of microscope most commonly seen in classrooms is the **compound light microscope**. This type of microscope has a light source found below the stage where a specimen is held in place on a glass slide. It also has two lenses held in place above the slide by a solid tube. The two lenses are like magnifying glasses that can be focused on the specimen. Together the two lenses compound the magnification. For example, the eyepiece, or ocular lens, may have a magnification of 10 times

(10×) and the second lens, or objective, may have a magnification of 40 times (40×). A specimen viewed through both lenses and the tube, with light from below the specimen clarifying the image, would be magnified 400 times (400×).

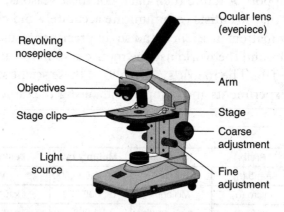

Revolving nosepiece

Objectives

Stage clips

Light source

Ocular lens (eyepiece)

Arm

Stage

Coarse adjustment

Fine adjustment

Figure 1–4. Example of a Compound Microscope

BRAIN TICKLERS   Set # 3

Select the correct term to complete each sentence.

1. In an experiment, the variable that represents the result is the (<u>dependent</u> or <u>independent</u>) variable.

2. In an experiment, the variable that represents the normal or unchanged state is the (<u>independent</u> or <u>control</u>) variable.

3. Microscopes most commonly available to students are called (<u>compound light</u> or <u>scanning light</u>) microscopes.

(Answers are on page 14.)

The **electron microscope** is another type of microscope used commonly in biology. Electron microscopes use magnetic fields to move electrons through an ultrathin slice of a specimen and create highly magnified images. These microscopes can provide an image of the structures that make up cells. A transmission electron microscope

(TEM) can magnify a specimen up to 500,000 times. A scanning electron microscope (SEM) focuses electrons to scan the surface of a sample to create a two-dimensional surface image and magnify it up to three million times. Clearly, microscopes are important technological devices for biologists who are trying to understand how all living things are put together, how they work, and how they interact.

 SUPER BRAIN TICKLERS

Match these definitions to the correct terms.

1. The microscope lens that your eye looks into

2. Statement describing a tested and observed predictable phenomenon

3. A microscope that uses electrons and magnetic fields to achieve magnifications up to 500,000×

4. A well-tested and supported explanation of events in the natural world

5. A variable that is changed in an experiment in order to produce results

6. A characteristic of all living things that allows them to maintain their internal balance

7. A characteristic of all living things that involves getting and using energy and nutrients and releasing waste

8. A characteristic of all living things that is seen in the way that all members of a species have their anatomy arranged in the same way

A. homeostasis

B. TEM

C. theory

D. organization

E. ocular

F. dependent

G. metabolism

H. law

I. independent

(Answers are on page 14.)

Vocabulary

Bias: When an investigator's preconceived ideas or beliefs influence the way he or she proceeds with research, an experiment, data collection, or data analysis.

Compound light microscope: A common light microscope that uses visible light and a series of lenses to produce magnified images of very small specimens.

Control variable: The condition in an experiment that does not change. It can be used to measure the change in a dependent variable.

Data: Information gathered from experimentation or research.

Dependent variable: An element of a scientific question that depends on something else; it is often the result of an experiment on an independent variable.

Electron microscope: A complex microscope that provides high magnification of very small objects using magnetic fields and electrons.

Graph: A diagram showing the relationship between variable quantities, which are set on an x-axis and a y-axis.

Homeostasis: Life process involving the adaptations in an organism that maintain its equilibrium, or balanced state, despite external conditions.

Hypothesis: A statement that predicts an outcome; it is used as a foundation of scientific investigation.

Independent variable: The element of a scientific investigation that is deliberately changed (resulting in the dependent variable).

Inference: A conclusion based on reasoning and evidence.

Law: A single fact of science that has been extensively tested, supported, and accepted. A scientific law tells us what will happen every time specific circumstances occur.

Metabolism: Life process when nutrients are taken in and broken down into small enough molecules to be used by the cells of a body; following the use of the molecules, waste products are eliminated.

Observation: The act of paying careful attention to the information obtained by our senses; all information is then recorded clearly, accurately, and honestly.

Organization: All living things are arranged in specific ways. There is a clear arrangement of organelles inside each cell. In multicellular organisms, cells, tissues, organs, and systems are found in very specific locations. Wouldn't it be strange if your fingernails were at the base of your fingers instead of at your fingertips?

Qualitative evidence: Scientific information gathered from sensory input: sounds, colors, textures, shapes, and odors.

Quantitative evidence: Scientific information collected in numerical form, as in graphs, tables, and charts.

Scientific method: A research process in which a question is asked, a problem is identified, relevant data are collected, a hypothesis is developed from these data, and the hypothesis is empirically tested.

Table: A visual collection and representation of data or information.

Theory: A scientifically accepted explanation of how nature works.

Variable: A factor in an experiment or investigation that is likely to change.

Brain Ticklers—The Answers

Set # 1, page 4

1. False
2. False
3. True

Set # 2, page 7

1. C
2. A
3. B

Set # 3, page 10

1. dependent variable
2. control variable
3. compound light

Super Brain Ticklers

1. E
2. H
3. B
4. C
5. I
6. A
7. G
8. D

Chapter 2

Cells

What Are Cells?

Cells are the basic structural units of all living things

So, do you remember from the last chapter that all living things are either cells, or are made of cells? And that cells are organized structures, with parts that perform specific and unique functions? This is important because cells are the structural units that make up all living things.

Cells are different sizes and shapes

Cells come in many sizes. Most, but not all, cells are microscopic, meaning that most individual cells can only be seen by using a microscope. A cell's size is limited by its surface area to volume ratio. This means that each cell needs to get nutrients and energy to its deepest, most interior regions, and nutrients get into the cell through its cell membrane. Therefore, the surface area of the cell membrane limits how much nutrition can enter the cell and become available to all the structures of the cell.

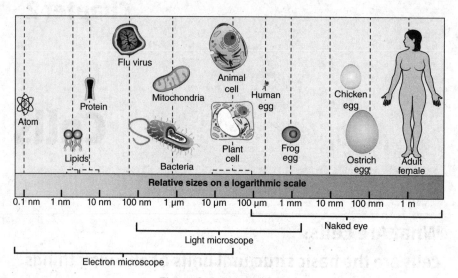

Figure 2–1. Relative Sizes

There are more than 200 different types of cells in the body of a complex organism like a human. These cells have many different shapes that make them most effective at performing their functions.

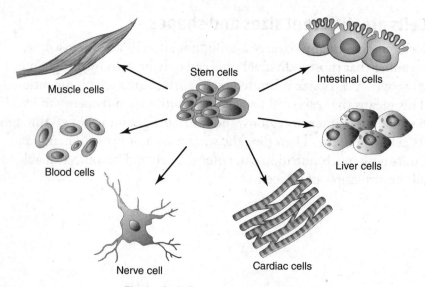

Figure 2–2. Types of Human Cells

Levels of Classification

The science of biological taxonomy names, identifies, and defines all known living things based on their shared characteristics and relationships with other living things. Living things are first identified as members of a domain (Archaea, Bacteria, or Eukarya), which are then divided into seven kingdoms (listed below under "**Diversity of Cells**"). Kingdoms are progressively split into smaller and more related groups in the following order: phylum > class > order > family > genus > species.

PAINLESS STUDY TIP

An mnemonic is a word or phrase that uses the first letter of each term you want to remember. Mnemonics are a great way to remember the order of topics. When you were younger, you may have learned "ROY G BIV" to remember the order of colors in a rainbow. Many science students use the mnemonic "Doting King Philip Came Over From Germany Saturday" to remember the levels of taxonomic naming. Try making up your own mnemonic for the levels of organization—it may be even easier for you to remember.

These scientific names allow scientists around the world to communicate accurately about the organisms with which they are working. When scientists refer to a very specific type of organism, they will use the genus and species names designated by taxonomy. For example, every domestic cat is scientifically named *Felis catus*. *Felis* is the genus name and *catus* is the species name. The naming rules require that the genus always begins with a capital letter and the species always begins with a lowercase letter. The genus and species should also be underlined or italicized. All living and extinct organisms that have been found have been named within this classification system. If a scientist wants to refer to multiple species of one genus, she may write the collective name as *Genus* spp. For example, if a scientist wanted to collectively discuss wolves (*Canis lupus*), coyotes (*Canis latrans*), and dogs (*Canis familiaris*), he could group the three together as *Canis* spp.

DOMAIN	Eukarya
KINGDOM	Animalia
PHYLUM	Chordata
CLASS	Mammalia
ORDER	Carnivora
FAMILY	Canidae
GENUS	*Canis*
SPECIES	*Canis lupus*

WOLF

Figure 2–3. Levels of Organization

PAINLESS TIP

Organization and classification of all life forms belongs to a science called "biological taxonomy." This branch of science was developed by Carl Linnaeus in the 1750s.

Linnaeus sorted and classified many plants and animals based on structural characteristics of their bodies and whether they lived primarily on land, in the water, or in the sky. Taxonomy is an ever-changing field that must continually be updated in order to classify newly discovered organisms and to incorporate the newest scientific information and methods.

Cells are prokaryotic or eukaryotic

The simplest cells are **prokaryotic** and are found in two domains: Archaea and Bacteria. These cells are very simple, and some are very similar to the first living things on Earth. Both domains are single celled, and so they perform all the functions of a living thing within their tiny, single-celled structures. Archaea are capable of surviving in extreme conditions, like extreme heat, salinity, and acidity, but are also found in milder environments. Archaea and Bacteria have an outer membrane to enclose their internal cytoplasm and structures, but they do not have membrane-bound organelles. Without membrane-bound organelles, they can't have a nucleus or mitochondria—their DNA simply floats in their cytoplasm. They often have an additional protective outer wall structure.

Cells of **eukaryotes** are more complex. Organisms of the third domain, Eukarya, called eukaryotes, can be either single celled or multicellular. Eukaryotes have both an external membrane and multiple internal membrane-bound organelles. Eukaryotic organelles include a nucleus, a membrane-bound structure that contains genetic material. The kingdoms contained in the domain Eukarya are Protista, Chromista, Fungi, Plantae, and Animalia.

Diversity of Cells

Kingdoms Archaea, Bacteria, Protista, Chromista, Fungi, Plantae, and Animalia

Each type of cell has different needs and performs different functions.

Archaeal and bacterial cells are prokaryotes found throughout the world. They conduct all the functions of any living cell, either on their own or in cooperation with other organisms. Both types of prokaryotes can reside inside and outside of other living bodies— usually, but not always, in a mutually beneficial relationship. Both archaea and bacteria commonly reproduce asexually by simple cell division, called **binary fission**.

> **PAINLESS TIP**
>
> The science of taxonomy has revised its classification system and criteria many times since Linnaeus's work. Domains and kingdoms have been added and the bases for classification have been modified as science and understanding of the world has improved. Current classification depends on genetics, cellular structure, nutrition mode, lifestyle, organization of body parts (morphology), and evolutionary history (phylogeny).

- Archaeal cells can live in common conditions but often live in extreme environments like volcanos, hot springs, salt lakes, and marshes, while bacteria are generally found in more common environments. Some types of archaea make their own nutrients from chemosynthesis (using other energy sources than sunlight) rather than photosynthesis (using sunlight).

- Bacterial cells live in every environment in the world. They are classified based on their shapes—a rod-shaped bacterium is called a bacillum, a spherical-shaped bacterium is called a coccus, and a spiral-shaped bacterium is called a spirillum. Bacteria are composed of cytoplasm surrounded by a cell membrane, a cell wall, and often by an exterior capsule. The cytoplasm usually contains a single circular strand of DNA and ribosomes. Some bacteria have hairlike structures that can help them move, stick to surfaces, or transfer their genetic information to other cells. Some bacteria that transmit human illnesses include *Vibrio cholera,* which causes cholera, and *Mycobacterium tuberculosis,* which causes tuberculosis (TB). Many bacterial diseases can be treated with drugs called antibiotics.

The simplest kingdoms of eukaryotes are the animal-like and fungus-like Protista and the plant-like Chromista. These organisms all contain cells that are complex, with membrane-bound organelles including the nucleus. They can be either single celled or multicellular. Cells of Protista and Chromista are both capable of reproducing either asexually or sexually.

- Protists are simple and **heterotrophic,** with cells needing to get their nourishment from outside their bodies, much like organisms in the Animalia and Fungi kingdoms. Protists need to be either motile (able to move like animals) or be capable of absorbing nutrients from their environment (like fungi). They contain at least one membrane-bound nucleus and other membrane-bound organelles. Most protists have mitochondria and cytoskeletal organelles like cilia and flagella to allow for movement. Amoebas are able to move by changing the shape of their cell membrane to use pseudopodia (false feet) to pull them along. A human infection with an amoebic protist called *Naegleria fowleri* (brain-eating amoeba) can be fatal when breathed in through the nose. African sleeping sickness and Chagas disease are caused by two different *Trypanosoma* species.

PAINLESS TIP

Chromista is the newest kingdom acknowledged by taxonomists. It was introduced in 1981 and has been gaining acceptance since the early 2000s. The Chromista separated from the Protista as a eukaryotic kingdom of simple plant-like organisms.

- Chromista are plant-like organisms with cells that are similar to plant cells. They are generally **autotrophs**, with organelles called plastids that contain a molecule called cytochrome c. This allows them to conduct photosynthesis, so they can manufacture their own food. Many Chromista also have tiny, external hair-like cytoskeletal structures called **cilia** and **flagella**. Originally, when they were split from the Protista, the Chromista kingdom only represented algae, but now it contains eight different phyla, including *Paramecium* and the dinoflagellates. Chromista also includes the parasites that cause malaria (*Plasmodium*), a brain-infecting parasite called *Toxoplasma*, and *Phytophthora* (which causes potato blight).

- Cells of Fungi are mostly similar to Animalia cells because both are heterotrophs and obtain their nutrients from preformed organic compounds. They do, though, have several unique features. Fungi have cellular structures that allow them to secrete enzymes to externally digest their food. They have cell walls (like plants) containing chitin (found in insect exoskeletons) rather than the cellulose found strengthening plant cell walls. Fungi often have multiple nuclei in a single cell. They may reproduce asexually or sexually. Examples of fungi include *Rhizopus stolonifer* (a bread mold), *Tinea pedis* (athlete's foot), and *Alternaria* spp. (some species of which can form mildew in the moisture of a bathroom).

- Plant cells, being autotrophic and photosynthetic, contain chloroplasts with chlorophyll. Most plants are multicellular and contain cells that support their photosynthetic activity, such as cells with strong cell walls, storage space for water, and mechanisms for seed dispersal. Many plant cells cycle between an asexual, sporophyte generation and a sexual, gametophyte/seed generation. Plants may be simple, like mosses (*Sphagnum* spp.) or horsetail ferns (*Equisetum* spp.), or they may be more complex, like eastern pine trees (*Pinus strobus*) or flowering plants (*Tulipa* spp.).

- Cells of animals, as heterotrophs, require structures that allow them to obtain their nutrients through preformed organic compounds. Animals also use a lot of energy in their attempts to either catch food or avoid becoming food, so they need structures in their cells that will allow them to store significant amounts of energy and move efficiently. Animals make offspring using sexual reproduction and make new cells for themselves using asexual reproduction. Most animals are classified as invertebrates or vertebrates. Invertebrates include jellyfish, like the Portuguese man o' war (*Physalia physalis*), earthworms (*Annelida* spp.); and beetles (*Coleoptera* spp.). Vertebrate animals include Atlantic salmon (*Salmo salar*), frogs (*Anura* spp.), ball pythons (*Python regius*), bald eagles (*Haliaeetus leucocephalus*), grizzly bears (*Ursus arctos*), and human beings (*Homo sapiens*).

BRAIN TICKLERS   Set # 1

Decide whether each of the following statements is true or false.

1. The size of cells is dependent on the number of organelles they contain.

2. Prokaryotic cells do not have a nucleus.

3. Only plants from the Plantae kingdom are able to use photosynthesis to produce the nutrients they need.

(Answers are on page 32.)

What Are the Parts of Cells?

Cells are made of *organelles*

Organelles are tiny structures found inside cells that help those cells to do their jobs. Each type of **organelle** performs specialized functions. Each organelle has a unique shape and structure that allows it to contribute to cellular function. Different types of cells have different numbers of each type of organelle so that they can perform the jobs they are required to do.

- The **cell membrane** is the part of a cell that keeps the outside out and the inside in. All cells have a cell membrane. Its chemical structure gives the cell membrane the ability to control what materials can enter and leave the cell. The cell membrane is made of a phospholipid bilayer made of two opposing layers of hydrophilic (water-attracting) heads and hydrophobic (water-repelling) tails, cholesterol molecules to maintain stability between the two layers, and embedded protein molecules to assist with molecular transport across the membrane.

- **Cytoplasm** is the gelatinous material that fills the cell membrane. It suspends organelles where they belong inside the cell and serves as a medium for many of the chemical reactions inside the cell. It contains a colorless gel-like substance called cytosol and is around 80 percent water. Cytoplasm is the material that suspends cellular organelles throughout the cell and is the location for the majority of chemical and cellular activities.

- The **nucleus** is a structure that contains most of the genetic material for the cell. The nucleus is surrounded by and separated from the cytoplasm by the **nuclear membrane**. **Nucleoplasm** is the gelatinous material that holds and suspends nuclear contents. Deep within the nucleus is the **nucleolus,** which serves to produce and assemble ribosomes.

- One of the smallest and most plentiful organelles in each cell is the **ribosome**. The job of the ribosome is to make proteins. Ribosomes are found throughout the cytoplasm but are mostly concentrated near the nucleus. Here they have easy access to the genetic instructions for the assembly of the proteins needed by the cell or the organism.

PAINLESS TIP

The most common organelle in eukaryotic cells is the ribosome. Ribosomes are also found in prokaryotic cells. Remember, all cells need to make proteins, so ribosomes are important!

- The **endoplasmic reticulum** is a system of interconnected tubules (made of membrane) that helps the cell to move materials around. There are two types of these tubules: rough endoplasmic reticulum (RER) and smooth endoplasmic reticulum (SER). Most RER is found close to the nucleus. The outside of the RER is covered with ribosomes, causing it to have a bumpy, or rough, surface. As the ribosomes on the RER surface manufacture proteins, the proteins enter the RER and are checked for accuracy, then transported to their next destination in the cell. The smooth ER is found farther from the nucleus and has a smooth outer surface, without ribosomes attached. The SER functions in lipid, glycogen, and steroid synthesis and transport.

- Cells need to make energy in order to survive and to do their jobs. Their **mitochondria** are the organelles that serve this function. Within the folded and layered double-membrane structure of the mitochondria, the molecule that provides energy to our cells is assembled. This molecule is called **adenosine triphosphate** (ATP), and in forming this molecule by attaching a third phosphate to an **adenosine diphosphate** (ADP) molecule, the cell can store small amounts of energy that help to drive cellular jobs and activities.

- Another organelle associated with energy is the **chloroplast**. The chloroplast is found in **autotrophs**, which are living things that use energy from the sun to chemically reassemble carbon dioxide (CO_2) and water (H_2O) into glucose ($C_6H_{12}O_6$) and oxygen (O_2). A chemical called chlorophyll, found in chloroplasts, traps solar energy to drive this reaction.

- A cell's **Golgi apparatus** is another tubular network of transport channels. It looks like a stack of underfilled water balloons and functions in the processing and packaging of proteins and lipids. Many of these molecules are packaged into sacs that carry them to the outer boundaries of the cell to be released from the cell and sent to be used by other cells in the body.

- The **lysosome** is an enzyme-filled, sac-like organelle that wraps around and engulfs waste materials in the cell. After wastes are dissolved, they are pushed out of the cell. Less often, lysosomes digest biomolecules that will ultimately be released and used inside the cell.

- The roles of the **cytoskeleton** include providing strength and structure to the cell. It is a dynamic and intricate system of protein fibers found in the cytoplasm of all cells, including prokaryotes. Internally, the cytoskeleton is made up of microtubules, microfilaments, and intermediate filaments. Externally, the cilia and flagella are cytoskeletal structures. Centrioles, structures that help align and direct chromosomes during cell division, are cytoskeletal. The cytoskeleton is also involved in holding and moving organelles and in many cell-signaling pathways.

- **Peroxisomes** are organelles whose main job is to break down hydrogen peroxide in all eukaryotic cells. Hydrogen peroxide (H_2O_2) is a byproduct of many cellular reactions; it becomes toxic if it accumulates in a cell.

- **Vacuoles** are water-filled, membrane-bound organelles that may contain enzymes as well as organic and inorganic molecules. They are found in all types of cells, including some bacterial cells. Their functions include holding water in plant cells, isolating waste and harmful materials in the cell, keeping internal plant cell pressure to avoid wilting, and holding up plant structures that might tend to pull down the top of a plant (like flowers and leaves).

- The **cell wall** is a tough and flexible layer covering the cell membrane. It is found in most prokaryotes, in some Protista, and in the Chromista, Fungi, and Plantae kingdoms. Animal cells do not have cell walls. The cell wall protects, supports, and regulates the internal pressure of the cell.

 CAUTION—Major Mistake Territory!

Cells that have a cell wall still have a cell membrane.

If a cell has a cell wall, it will surround the cell membrane to provide strength and structure to the cell.

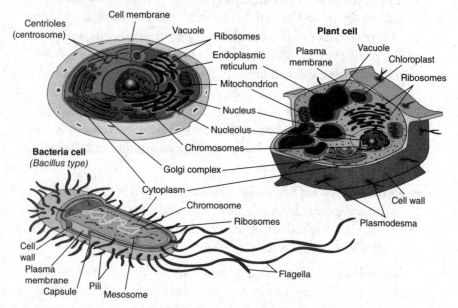

Figure 2–4. Some Typical Cells

PAINLESS STUDY TIP

Always spend time reviewing the images found in the text. Try to relate the words in the section to the images you see. Try drawing and labeling the images for yourself. These techniques can help you clarify your questions, embed the information in your brain, and help you to identify relationships within the image.

BRAIN TICKLERS   Set # 2

Name the term that represents each of the following descriptions.

1. Cellular structures that assemble proteins

2. Eukaryotic structures that store and protect the cell's DNA

3. Structure that engulfs and digests cellular waste to prepare for disposal

A. cell membrane

B. endoplasmic reticulum

4. Tubular networks in the cell that package and transport materials inside and to the outside of the cell

5. Outer boundary of all cells; it is able to select what molecules can enter and exit the cell

C. nucleus

D. ribosomes

E. lysosome

(Answers are on page 32.)

What Happens When Cells Don't Work Correctly?

Organelle diseases

Any cell and any organelle can stop working correctly. Most often, those organelles or cells will be destroyed and eliminated, and little to no harm is done. Sometimes, though, the genetic instructions to build the organelle are incorrect and this can cause serious problems. Cystic fibrosis, for example, is a genetic disorder that can cause severe respiratory and digestive issues. Cystic fibrosis is a result of misshapen cell membrane proteins. Mitochondrial diseases can reduce available cellular energy and can affect the muscular and nervous systems. Lysosomal storage diseases can cause lethal accumulations of cellular waste products in diseases like Tay-Sachs.

Cancer

Cancer is essentially "cells gone wild." Cancer cells do not obey the rules they are supposed to follow—particularly the rules about when to grow and divide, and when not to. This "misbehavior" leads to increased rates of cell division or a decrease in **apoptosis,** or genetically programmed cell death. Apoptosis is actually a normal and healthy process for a multicellular organism like us, removing damaged cells and keeping cells in the appropriate number in body tissues. Cancer can result from normal cellular controls being suppressed. Commonly, issues like a loss of cell-to-cell contact inhibition, which normally causes cells to stop growing when they are touching another cell, can then develop. As these cells grow out of control, tumors may form and spread and may ultimately become cancerous.

BRAIN TICKLERS   Set # 3

Select the correct term to complete each sentence.

1. Any time DNA, an organelle, or a cell stops working correctly it, (<u>will</u> or <u>may or may not</u>) cause serious complications.

2. Potentially deadly collections of cellular waste will occur in diseases that affect the (<u>lysosomes</u> or <u>mitochondria</u>).

3. Cells are programmed to die after a specific number of replications or at a certain age. This programmed cell death is also known as (<u>aging out</u> or <u>apoptosis</u>).

(Answers are on page 32.)

SUPER BRAIN TICKLERS

Match these definitions to the correct terms.

1. The tough, exterior cellular structure that allows plants to grow tall and strong in spite of gravity's pull toward Earth

2. A molecule that provides cellular energy in its high energy, third phosphorus bond

3. Asexual reproduction seen in bacteria

4. A domain that includes complex organisms with cells that contain nuclei

5. A chemical process that allows some organisms to produce nutrients for themselves

6. Organisms that cannot manufacture their own nutrients. They must search out and digest their food

A. photosynthesis

B. binary fission

C. prokaryote

D. cell wall

E. autotroph

F. eukaryote

G. ATP

H. heterotroph

7. Organisms that can manufacture food for themselves, often using photosynthesis

8. A domain containing the simplest organisms. They do not have a nucleus to contain their genetic material

(Answers are on page 32.)

Vocabulary

Adenosine diphosphate (ADP): A stable biological molecule made of adenosine and two phosphate groups. It is always ready to accept a third phosphate and to store energy in the bond holding that phosphate.

Adenosine triphosphate (ATP): An unstable biological molecule made of adenosine and three phosphate groups. It is the main molecule that provides the necessary energy to keep our cells working.

Apoptosis: Genetically programmed cell suicide.

Autotroph: A producer; an organism that makes its own food through photosynthesis or chemosynthesis.

Binary fission: Asexual reproduction involving a parent cell splitting into two identical daughter cells.

Cell membrane: The cell membrane; structure surrounding a cell's cytoplasm. It selects what enters and exits the cell.

Cell wall: Gives strength and support to cells. Tough, cellulose-based material that surrounds the cell membranes of plants, some fungi, and some protists.

Chloroplast: The organelle in photosynthetic cells that traps energy from sunlight; it contains structures that convert light energy to chemical energy.

Cilia: Short, hairlike structures of the cell's external cytoskeleton. They assist in obtaining nutrients and in locomotion.

Cytoplasm: A gelatin-like material; it comprises most of the substance of the cell, suspends organelles, and provides a medium for the cell's chemical reactions.

Cytoskeleton: A group of structures, including microtubules, microfilaments, cilia, and flagella, that give support, mobility, and strength to the cell.

Endoplasmic reticulum: A network of tubules that move materials inside the cell. There are two types of ER: smooth (SER) and rough (RER).

Eukaryotes: Cells that have complex membranes to contain their organelles, particularly important in enclosing DNA inside a nucleus. All living things other than bacteria are eukaryotic.

Flagella: Long hair-like exoskeletal structures, often used for movement.

Golgi apparatus: Stacked structures that sort and package molecules inside the cell into secretory vesicles for distribution both inside and outside the cell.

Heterotroph: A consumer; an organism that cannot make its own food and must consume other living things for nourishment.

Lysosome: An organelle that holds and digests nutrients for the cell and then discards the waste, usually outside the cell.

Mitochondria: Organelles that convert sugar's energy into chemical energy, which is then stored in the bond of the third phosphate group in ATP.

Nuclear membrane: A porous structure surrounding the cell's nucleoplasm.

Nucleolus: Organelle inside the nucleus where ribosomes are made.

Nucleoplasm: Cytoplasm inside the nucleus, contained by the nuclear membrane.

Nucleus: Organelle that contains chromosomes and nucleolus. It is found only in eukaryotes.

Organelle: A structure in a cell that performs a unique function.

Peroxisome: Organelle that holds enzymes that remove peroxide from cells, helps to break down fatty acids, and helps the cell manufacture cholesterol, myelin, and bile acids.

Prokaryotes: Organisms with a simple membrane system and no nucleus, so DNA is loose in the cytoplasm. All bacterial life is prokaryotic.

Ribosome: Small organelle that builds amino acids into proteins according to mRNA instructions.

Vacuole: Water-filled organelle that stores molecules and eliminates waste from cells. Vacuoles are larger and more common in plant cells, helping plants to provide internal pressure against plant cell walls (preventing them from wilting).

Brain Ticklers—The Answers

Set # 1, page 22
1. False
2. True
3. False

Set # 2, pages 26–27
1. D
2. C
3. E
4. B
5. A

Set # 3, page 28
1. may or may not
2. lysosomes
3. apoptosis

Super Brain Ticklers
1. D
2. G
3. B
4. F
5. A
6. H
7. E
8. C

Chapter 3

Biological Chemistry

What Are Atoms?

Atoms are the smallest units of matter, making up both organic (living) and inorganic (nonliving) materials. They form chemical elements and make up every solid, liquid, and gas. Atoms are made up of structures called protons and neutrons, located in the nucleus of the atom; and electrons, orbiting in levels surrounding the nucleus. Protons are positively charged, neutrons have no electrical charge, and electrons are negatively charged.

Elements are found on the periodic table of the elements; they are listed on this table based on their atomic number, which is the number of protons found in each atom's nucleus. All atoms have an equal number of protons and electrons, and they do not carry an electrical charge. Atoms can sometimes gain or lose electrons in their interactions with other atoms, and when the number of protons and electrons is not equal, an atom then becomes an ion.

PAINLESS TIP

Atoms are made up of mostly empty space. The atomic nucleus contains almost all of the atom's mass, and it, too, has a lot of empty space. If an atom were the size of a city block, the nucleus would be smaller than a pea at the center of that block!

PAINLESS TIP

To better understand the relationships between types of elements, review the periodic table (below). You'll find that most of the elements needed by living things have lower atomic numbers and are smaller and simpler atoms.

Figure 3–1. Periodic Table of the Elements

In living things, the following elements are considered essential and together make up over 97 percent of all living organisms: carbon (C), hydrogen (H), oxygen (O), nitrogen (N), phosphorus (P), and sulfur (S). Additionally, several elements are considered essential trace elements; these include calcium (Ca), potassium (K), sodium (Na), chlorine (Cl), and magnesium (Mg). Other elements are needed for very specific jobs and are used in very small amounts, including iron (Fe), zinc (Zn), and iodine (I).

PAINLESS TIP

The mnemonic CHONPS can help you remember the macronutrient elements that are found in abundance in living things. It tells you that carbon, hydrogen, oxygen, nitrogen, phosphorus, and sulfur are the most common elements that make us up.

What Are Molecules?

Molecules are the smallest particles of a substance, formed by groups of two or more atoms held together by chemical forces. Molecules are neutral, having no electrical charge. Molecules can be made up of a single element, like oxygen (O_2), or they can be made up of different types of elements bonded together, like water (H_2O). Table salt is a molecule made from sodium (Na) and chloride (Cl); its chemical formula is NaCl. Table sugar has a formula of $C_6H_{12}O_6$. In those molecules that contain small numbers, called subscripts, the numbers tell you how many atoms of the element are present in one molecule. So, one molecule of simple sugar would contain six atoms of carbon, twelve atoms of hydrogen, and six atoms of oxygen.

Sometimes these molecules are combined into formulas. A formula will tell you which molecules are being combined (the **reactants**) and what they will produce (the **products**). Often these reactions will contain different amounts of the reactant and product molecules. The formula would then look like this: $6CO_2 + 6H_2O \rightarrow C_6H_{12}O_6 + 6O_2$. This would read as 6 carbon dioxide molecules plus 6 water molecules yields 1 sugar molecule plus 6 oxygen molecules. This reaction describes photosynthesis and would occur in the presence of sunlight. As you read the formula, you can see that there is a full-sized number preceding each single molecule. That preceding number tells you how many molecules each are needed to make this reaction work.

BRAIN TICKLERS   Set # 1

Select the correct term to complete each sentence.

1. Elements are listed on the periodic table based on their atomic (<u>mass</u> or <u>number</u>).

2. The main elements found in all living things are (<u>C, H, O, N, P, S</u> or <u>C, H, O, Na, Cl</u>).

3. The subscript numbers in a molecular formula, like the 2 in H_2O, indicate how many (<u>atoms or reactants</u>) there are in a particular molecule.

(Answers are on page 51.)

What Are Chemical Bonds?

A chemical bond is a binding connection between atoms based on either the sharing or the exchange of electrons. Chemical bonds create molecules by holding together two or more atoms of the same or of different elements.

Covalent bonds: A covalent bond is formed when electrons are shared between two or more atoms. The shared electrons produce an attractive force that creates a stable bond between the atoms, which in turn create a relatively stable molecule like water or sugar ($C_6H_{12}O_6$).

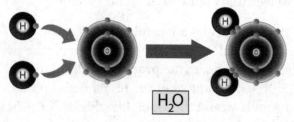

Figure 3–2. Covalent Bond

Ionic bonds: Ionic bonds are formed when electrons from the outermost electron level of atoms are given up by one atom and are taken in by another atom. This exchange provides each atom with a full and stable outer electron level. It also produces an electrical attraction between the two atoms. If the atom has more electrons than protons, it will be a negatively charged **ion**. If the atom has more protons than electrons, it will be a positively charged ion. An example would be salt (NaCl), which is composed of sodium (Na+) electrically attracted to chloride (Cl-). These two charged atoms, or ions, will hold together because of their electrical attraction.

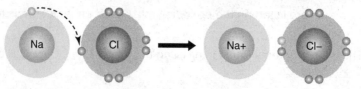

Figure 3–3. Ionic Bonding
(an electron is given up by sodium
and taken in by chloride)

Peptide bonds: Peptide bonds form proteins by holding **amino acids** together in a very specific sequence to make the protein correctly. A protein can be made of anywhere from 300 to thousands of amino acids held together by peptide bonds.

Hydrogen bonds: Hydrogen bonds are found in many of the molecules of living things. One important function of hydrogen bonds is to hold together the nitrogen base pairs in DNA.

Why do chemical bonds matter in biology?

One of the shared characteristics of all living things is **metabolism**. Metabolism refers to all of the chemical reactions that occur in the body of any living thing. These chemical reactions relate to the cell's energy, either making, storing, or using energy, to drive reactions in the cell. Metabolism also refers to the processes in a cell that involve repair and growth. Metabolic processes create or break chemical bonds. The metabolic processes are **catabolism** and **anabolism**. Anabolic pathways use energy and create bonds to build larger molecules from smaller building blocks. Catabolic pathways release energy as they break larger molecules apart into smaller component parts.

BRAIN TICKLERS   Set # 2

Match these descriptions to the correct terms.

1. Binds amino acids together   **A.** hydrogen bonds

2. Electrons are shared   **B.** ionic bonds

3. Binds nitrogen base pairs in DNA   **C.** peptide bonds

4. Electrons are given up/taken in   **D.** covalent bonds

(Answers are on page 51.)

Water and pH

Why is water so important in living things?

All living things require water to survive for a number of reasons. First, all living things are made up of around 65–70 percent water.

Water acts as the living environment for many organisms and as a helper to the process of respiration for organisms that are dependent on oxygen. Water acts as a pH and a temperature buffer, maintaining homeostasis and a healthy cellular environment. Water acts as a solvent, as a transporter of materials in the body, and as a cushion and shock absorber. Water pressure in plants helps them maintain their shape and internal structure. The elements in water drive many metabolic processes. Anabolic metabolic pathways use water and energy to help build larger molecules during **dehydration synthesis**. The opposite process, **hydrolysis**, is a catabolic pathway that takes larger molecules apart, releasing water and energy.

CAUTION—Major Mistake Territory!

It can be easy to mix up dehydration synthesis with hydrolysis. It can help to break the words down.

Dehydration should be a familiar term; it means the loss of water. Synthesis means to build. So, dehydration synthesis refers to the process of putting together two molecules (synthesizing) by removing a molecule of water (dehydrating).

Hydrolysis breaks down to *hydro* (water) and *lysis* (splitting apart). Hydrolysis refers to water being inserted into a molecule in order to break it apart.

Dehydration Synthesis

Hydrolysis (Digestion)

Figure 3–4. Dehydration Synthesis and Hydrolysis (Digestion)

What is pH and why does it matter?

When water splits into two ions, the ions are hydrogen (H+) and hydroxide (OH-). As the concentration of hydrogen ions increases, the water becomes more acidic. As the concentration of hydrogen ions decreases, the water becomes more basic (or alkaline). Measuring the pH of a liquid tells us how acidic or basic that liquid is. The pH scale goes from 0 to 14; pure water is in the middle at 7.0 and is called neutral. As the pH moves from 6.9 to 0, the liquid becomes exponentially more acidic. As the pH goes from 7.1 to 14, the fluid becomes exponentially more basic.

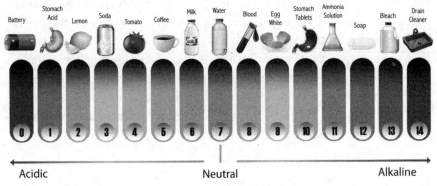

Figure 3–5. The pH Scale

PAINLESS TIP

The pH in a living thing is important because proteins in the organism can become denatured (nonfunctional and misshapen) and will no longer work if the pH goes out of the normal range. This will result in the death of the organism.

Biological Macromolecules

What is a biological macromolecule and how is it different from other molecules?

There are many kinds of molecules in the world, but most of the major types of molecules that make up living things belong to one of four categories: **carbohydrates**, **lipids**, **proteins**, and **nucleic acids**.

These four types of biological macromolecules, or biomolecules, have some special properties that make them uniquely able to build and maintain all types of living things. These are often large molecules, called **polymers**, composed of many smaller subunits, called **monomers**.

What is important to know about carbohydrates?

Carbohydrates, also known as sugars and starches, are used to help make energy available to cells. The elements found in carbohydrates are carbon, hydrogen, and oxygen. Tiny single-molecule sugars are like building blocks that give cells quick, easily accessible energy, while large carbohydrate molecules give the cells a stored form and a larger amount of energy that is not as easily obtained by the cells. Simple sugars with a single, unbonded sugar molecule are called monosaccharides; two sugar molecules bonded together are called disaccharides; and when multiple sugar molecules are bonded together, they make a polysaccharide. Simple carbohydrates are found in fresh fruits and vegetables. Complex carbohydrates are present in grains, potatoes, rice, pasta, and bread.

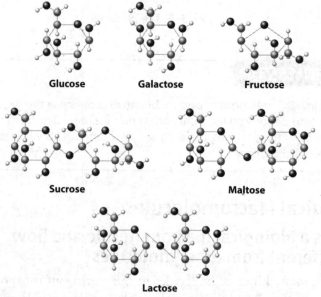

Figure 3–6. Carbohydrates

What is important to know about lipids?

Lipids are also known as fats. They are large molecules made up of carbon, hydrogen, and oxygen; but lipids have many more hydrogen atoms than oxygen atoms. Like complex carbohydrates, lipids are molecules the body can use to store energy. The building-block molecules of lipids include fatty acids and, often, glycerol. Types of lipids include oils, waxes, and steroids. Lipids do not dissolve in water. Butter, margarine, many meats, cream, and mayonnaise are all dietary sources of lipids.

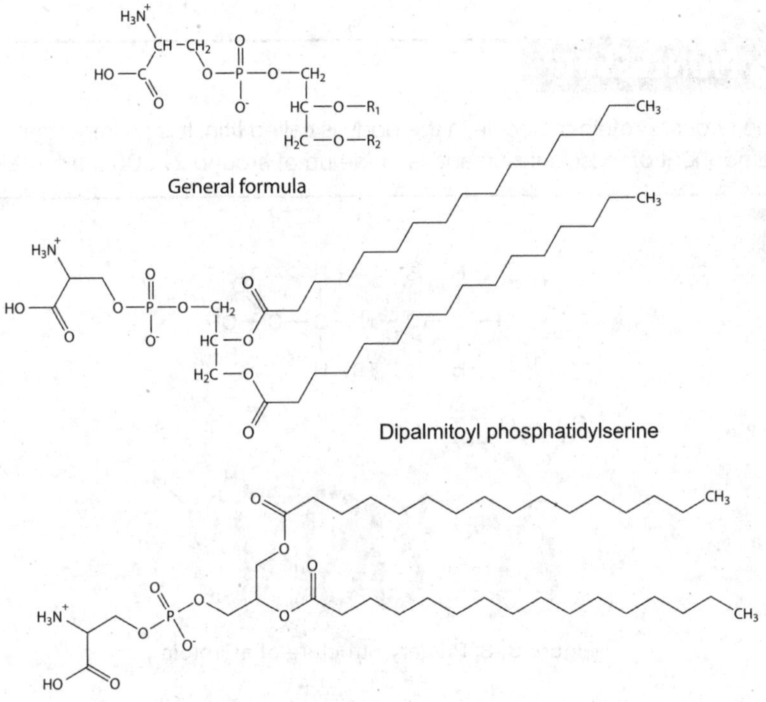

General formula

Dipalmitoyl phosphatidylserine

Figure 3–7. Lipids: Phosphatidylserine

What is important to know about proteins?

Proteins are important molecules that do a number of jobs. Proteins form the structures and scaffolds that make up all living things, help with cellular and molecular transport, act as hormones and antibodies, and assist with chemical reactions. They are made of carbon, hydrogen, oxygen, and nitrogen. Proteins are assembled with building blocks called amino acids held together by peptide bonds. Dietary sources of proteins include meat, eggs, legumes, and dairy products.

There are only 20 amino acids, but because they can combine in many different ways, it's estimated that they can form more than 80,000 different proteins. Human bodies manufacture eleven of these amino acids; these are called "nonessential" amino acids. The nine amino acids that humans cannot produce are called "essential" amino acids. Essential amino acids must be consumed in the human diet in order to become available to be incorporated into proteins.

PAINLESS TIP

The largest protein molecule in the body is called titin. It is an important component of muscle tissue and is made up of around 27,000 amino acids.

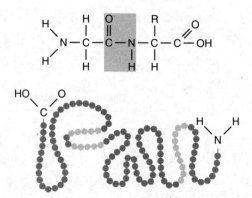

Figure 3–8. Primary Structure of a Protein

What is important to know about nucleic acids?

Nucleic acids are the molecular structures that carry our hereditary information. There are two types of nucleic acid—DNA and RNA. **Deoxyribonucleic acid**, or DNA, makes our chromosomes, which are like a cookbook with instructions to make each of us. DNA is generally found in the cell's nucleus. Ribonucleic acid, RNA, does the work to follow the instructions in DNA in order to make living cells and organisms. The elements found in nucleic acids are carbon, hydrogen, oxygen, nitrogen, and phosphorus. These elements are arranged into nucleotides, the building blocks of DNA and RNA.

The structure of DNA is called a double helix. This means that it has two backbones, or side rail strands, held together by nitrogen base pairs that look like the rungs on a ladder. The two side rails are made up of identical sequences of sugar (deoxyribose) and a phosphate group. These repeat over and over again for the entire length of the DNA strand. The only part that is different in the strand are the rungs—the nitrogen base pairs. The nitrogen bases of DNA are attached to the deoxyribose and include adenine (A), thymine (T), cytosine (C), and guanine (G). These nitrogen bases are paired very specifically in order to give the code, or recipe, to assemble the correct protein. Adenine always bonds with thymine with a double hydrogen bond. Cytosine always bonds with guanine using a triple hydrogen bond. The sequence of adenine, thymine, cytosine, and guanine in the DNA strand indicates the sequence of amino acids to be assembled during transcription and translation (see Chapter 10—Molecular Genetics).

Ribonucleic acid, or RNA, is a single-stranded molecule. Its backbone is made up of ribose (a different sugar from DNA) and the same phosphate group that DNA uses. Attached to each ribose is a nitrogen base: adenine, cytosine, guanine, and uracil. Uracil (U) replaces thymine in RNA. There are three kinds of RNA: transfer RNA (tRNA), ribosomal RNA (rRNA), and messenger RNA (mRNA). The job of tRNA is to bring the correct amino acid to complement the mRNA and build the new protein. The rRNA is used as a part of the ribosome where proteins are built. The job of mRNA is to make a complementary copy of the DNA during transcription. This means that when the nitrogen base in the DNA is adenine, the mRNA nitrogen base will be uracil; when the DNA base is thymine, mRNA will match up with an adenine; the DNA base guanine will bond with mRNA cytosine; and DNA cytosine will bond with mRNA guanine. The mRNA code will ultimately direct the assembly of a series of amino acids to become a protein.

PAINLESS TIP

The DNA strand A T C G C A T A G will transcribe to this

mRNA strand: U A G C G U A U C

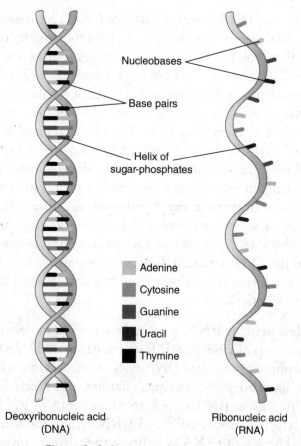

Adenine

Cytosine

Guanine

Uracil

Thymine

Deoxyribonucleic acid
(DNA)

Ribonucleic acid
(RNA)

Figure 3–9. Structure of DNA and RNA

BRAIN TICKLERS   Set # 3

Match these descriptions to the correct terms.

1. Provide stored energy and steroids   **A.** proteins

2. Can give cells quick or stored energy   **B.** lipids

3. Carry hereditary information   **C.** nucleic acids

4. Serve as enzymes and structural components   **D.** carbohydrates

(Answers are on page 51.)

Specialized Biological Molecules

What is an enzyme?

An **enzyme** is a special type of protein that reduces the **activation energy** required for a cell to initiate a chemical reaction. This means that a chemical reaction that, in a lab, might require additional energy such as heat or mixing can occur in a living organism without needing to be heated or mixed. This reaction can happen because an enzyme provides the push needed to drive the reaction.

PAINLESS STUDY TIP

Many enzymes end with the suffix -*ase*. For example, the milk that you drink contains a milk sugar called lactose. The enzyme in your body that helps you digest lactose is called lactase. If you lack the enzyme lactase, taking in milk products will probably upset your stomach. Another digestive enzyme called amylase helps us to **catalyze**, or break down, the carbohydrates, or sugars and starches, that are present in our mouths when we begin to eat.

Enzymes work by joining with a **substrate** in what is called an **induced-fit model**. In an induced-fit model there are two parts, a substrate and an active site, that join together and then change shape in order to initiate the process of catalyzing the reaction. When the reaction is complete, the substrate and active site separate, and the active site can catalyze its next reaction.

PAINLESS TIP

If you add hydrogen peroxide to a piece of uncooked liver, the peroxide will bubble up, water will be produced, and there will be an increase in temperature. This is because the liver contains an enzyme, catalase, to remove toxic hydrogen peroxide waste from our bodies. In this experiment, the catalase in the liver reacts with the hydrogen peroxide (H_2O_2) to form water (H_2O) and oxygen (O_2), which is seen as bubbles.

ENZYME

Induced Fit Hypothesis

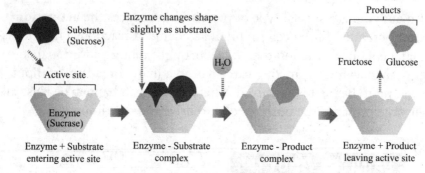

Figure 3–10. Structure of an Enzyme

Why are enzymes important?

Life would not exist without enzymes. They allow reactions that otherwise would not occur without some physical exertion that wouldn't be possible in a living organism, such as heating or mixing. Enzymes drive, or catalyze, many types of chemical reactions in all living things. These reactions include digestion and metabolism, respiration and breathing, reproduction, growth, disease processes, healing, and maintenance of a healthy organism.

Adenosine Triphosphate (ATP)

What is ATP?

Adenosine triphosphate (ATP) is an important molecule in living things because it provides energy to drive reactions. The high energy bond between adenosine diphosphate (ADP) and a third phosphate group (Pi) is very unstable, meaning it does not require much energy to break this third bond. Once the bond is broken, a small amount of energy is released and the ADP is once again available to bond with another phosphate molecule and the (now free-floating) phosphate becomes available to bond with another ADP molecule. This molecular relationship is constantly recycled in all living things as low-energy ADP bonds with phosphate to form high-energy ATP.

Why is ATP important?

The energy provided in the hydrolysis of ATP (to reform ADP plus Pi) is used in many essential processes in cells and living organisms. ATP hydrolysis provides energy to allow active transport across cell membranes, synthesis of DNA and RNA, cell signaling both inside and between cells, muscle contractions, and other cellular functions.

BRAIN TICKLERS   Set # 4

Select the correct term to complete each sentence.

1. One major source of energy for cells is (<u>ADP</u> or <u>ATP</u>).

2. Enzymes or substrates are recycled back into use when they're done with each reaction.

3. Enzymes reduce (<u>kinetic</u> or <u>activation</u>) energy in order to drive reactions in living organisms.

(Answers are on page 51.)

SUPER BRAIN TICKLERS

Match these definitions to the correct terms.

1. A bond in which an electron is taken from one ion and given to another

2. The acid-base balance created when water breaks apart

3. Created when amino acids are bonded together

4. Often made of glycerol and fatty acids

5. DNA and RNA

6. Allows chemical reactions to occur at normal body temperatures

A. enzyme

B. nucleic acid

C. ionic bond

D. atomic number

E. carbohydrate

F. covalent bond

G. pH

H. protein

7. Determines an element's position on the   I. lipid
   periodic table

8. A bond in which are shared between
   elements

(Answers are on pages 51–52.)

Vocabulary

Activation energy: Extra energy from an enzyme that provides the push needed to drive a reaction.

Adenosine triphosphate (ATP): A molecule in living things that provides energy to drive reactions.

Amino acid: The molecular building block of a protein.

Anabolism: Reactions that use energy and create bonds to build larger molecules from smaller building blocks.

Atom: The smallest unit of matter, making up both organic (living) and inorganic (nonliving) materials.

Carbohydrates: Biomolecules also known as sugars and starches; used to help make energy available to cells. The elements found in carbohydrates are carbon, hydrogen, and oxygen.

Catabolism: Reactions that break down large molecules.

Covalent bond: Formed when electrons are shared between two, or more, atoms.

Dehydration synthesis: A process during which anabolic metabolic pathways use water and energy to help build larger molecules.

Deoxyribonucleic acid (DNA): The main storage molecule of hereditary information; DNA is a double helix nucleic acid that makes up each chromosome.

Enzyme: Protein based molecule that reduces the activation energy needed to drive a reaction.

Hydrolysis: A catabolic pathway that takes larger molecules apart, releasing water and energy.

Induced-fit model: A model of enzyme activity consisting of two parts: a substrate and an active site that join together and then change shape in order to begin the process of catalyzing the reaction. When the reaction is complete, the substrate and active site separate, and the active site can catalyze its next reaction.

Ion: Formed when an atom has an unequal number of electrons and protons.

Ionic bond: Formed when electrons from the outermost electron level of atoms are given up by one atom and are taken in by another atom.

Lipids: Biomolecules also known as fats; large energy-storage molecules.

Metabolism: All of the chemical reactions that occur in the body of any living thing. These chemical reactions relate to the cell's energy—either making, storing, or using energy; driving reactions within the cell; or running processes that involve cellular repair and growth.

Molecule: The smallest particle of a substance, formed by a group of two or more atoms held together by chemical forces.

Monomers: Small molecules that may be combined to form polymers. In living things, monomers are generally small enough to be absorbed into cells.

Nucleic acids: The molecular structures that carry our hereditary information.

Polymers: Large molecules composed of many smaller subunits, called monomers.

Products: Molecules that are produced after reactants are combined.

Proteins: Biomolecules, made of amino acids, that form the structures and scaffolds that make up all living things, help with cellular and molecular transport, act as hormones and antibodies, and assist chemical reactions.

Reactants: Molecules that are being combined to produce a product.

Ribonucleic acid (RNA): A single-stranded nucleic acid that helps DNA in the transfer of genetic information and in the building of proteins directed by that information.

Substrate: The substance acted upon by an enzyme.

Brain Ticklers—The Answers

Set # 1, page 35
1. number
2. CHONPS
3. atoms

Set # 2, page 37
1. C
2. D
3. A
4. B

Set # 3, page 44
1. B
2. D
3. C
4. A

Set # 4, page 47
1. ATP
2. Enzymes
3. activation

Super Brain Ticklers
1. C
2. G
3. H

4. I
5. B
6. A
7. D
8. F

Chapter 4

Cellular Transport

What Is Cellular Transport?

All living cells are surrounded by a cell membrane. This cell membrane is built similarly in all cells. Every cell must engage in transport across these cell membranes in order to stay alive. Remember, all living things conduct metabolism to get the energy and materials they need to stay alive. They must then have a way to get rid of unneeded waste materials. Cells also need other molecules to be exchanged depending upon the type of cell and its function. The way that molecules and materials are exchanged between the inside and outside of a cell is called cellular transport. Transport inside a cell is called **intracellular transport**, and transport between different, often adjoining, cells is called **intercellular transport**.

Eukaryotic cells use a similar type of membrane to surround some of their organelles. The mitochondria, nucleus, and chloroplasts are all surrounded by a lipid bilayer membrane. Other organelles have single lipid layers separating their organelle contents from the surrounding cytoplasm. Organelles also have different proteins and other embedded molecules from those in the cell membrane in order to perform their unique functions.

Cell Membrane Structure

How does cell membrane structure affect its function?

The cell membrane is arranged in a phospholipid bilayer. Phospholipid means that it's made of phosphate groups attached to lipid (fat) molecules and bilayer means it's arranged into two layers. The phosphate groups found in the cell membrane are arranged in a pattern similar to what you would see in a water-filled tub with a layer of ping-pong balls floating on top and all touching each other. In the cell, or plasma, membrane, these spherical structures are called polar heads. This means they have an imbalanced, or partial, charge that allows them to attract water molecules, making them **hydrophilic**. Finally, a pair of fatty acid tails extends from the bottom of each polar head and is aimed toward the opposite phospholipid layer (Figure 4–1) . Fatty acid tails are **hydrophobic** and have a polarity that repels water.

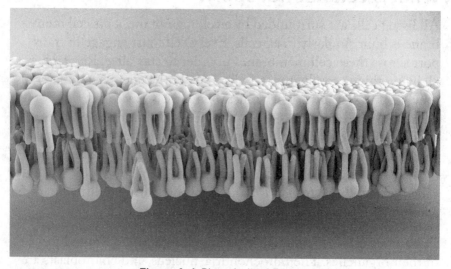

Figure 4–1. Phospholipid Bilayer

If the phospholipid bilayer was the only thing making up the cell membrane, the inner and outer layers could easily move in different directions and might trap important molecules in between the layers. Instead, the inner and outer layers are held together and are stabilized by cholesterol molecules. These sturdy cholesterol molecules also help protect the cell from extremes of temperature.

In addition, proteins are important embedded structures in cell membranes. There are two main categories of membrane proteins: those that cross the entire cell membrane are called integral proteins, and those that are attached to either the inner or outer phospholipid layer and only cross halfway into the bilayer are called peripheral membrane proteins. Proteins serve a number of different functions in cell membranes. Many of the proteins in a cell membrane act as transport channels to allow movement of specific molecules through the cell membrane.

REMINDER

Remember the biomolecules we explored in Chapter 3? We've talked a lot about proteins and lipids, right? Carbohydrates are also found attached to the outer cell membrane surface. When they are bound to lipids, they are called glycolipids. When attached to proteins, they are called glycoproteins. Both of these molecules can help the cell interact with other cells or can act as chemical receptors, or markers, for cellular recognition. Glycoproteins can help the cell with enhanced disease protection, enzyme production, and other functions, including the clotting of blood.

All of these molecules work together to form the barrier between the outside and inside of a cell. They also work together to determine which molecules are able to cross over from one side of the cell membrane to the other; this ability is called **selective permeability** (Figure 4–2). Because all of these pieces are able to stay in constant "fluid" motion and the parts that make up the outside layers of the cell membrane look like an artistic mosaic, scientists call their representation of the cell membrane the **fluid mosaic model**.

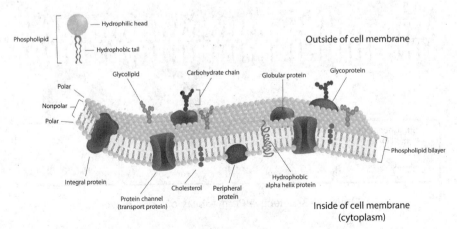

Figure 4–2. Structure of the Selectively Permeable Membrane

How is transport controlled by the cell membrane?

The selective permeability of every living cell membrane determines the molecules and ions that will move into and out of each cell to keep it alive. This movement is based on the size and electrical charge of the materials trying to cross the membrane. Tiny molecules and those that are small and polar (but uncharged) have the best chance of squeezing through between the spherical phospholipid molecules easily. Larger molecules and ions can't squeeze through, so they require help to cross into and out of cells.

BRAIN TICKLERS   Set # 1

Match these descriptions to the correct terms.

1. A membrane that allows only some particles across

2. Molecules with an imbalanced charge

3. A representation of the plasma membrane that represents how the layers move against each other

A. fluid mosaic model

B. selectively permeable membrane

C. polarity

(Answers are on page 67.)

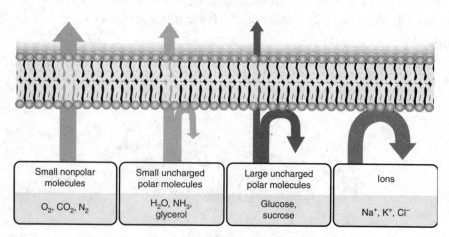

Small nonpolar molecules	Small uncharged polar molecules	Large uncharged polar molecules	Ions
O_2, CO_2, N_2	H_2O, NH_3, glycerol	Glucose, sucrose	Na^+, K^+, Cl^-

Figure 4–3. Selective Permeability of Lipid Bilayer

What Is Passive Transport?

When a larger molecule or an ion needs to enter or exit a cell, it is likely to need some assistance. There are a number of options for getting this help, and some form of **passive transport** is going to be the preferred choice. The term *passive* indicates that no energy is required to enable this type of chemical movement. All types of passive transport fall under the category of **diffusion**.

What is diffusion?

Diffusion is a form of passive transport where molecules move from an area where they are highly concentrated, spreading out until they are less concentrated and spreading out equally within the area. Simple diffusion occurs when the molecules are small enough to squeeze through the cell membrane, able to move just because there is a concentration gradient. When molecules are evenly distributed, they are in a state of **dynamic equilibrium**. Cells don't need to exert any energy to move molecules when those molecules are bouncing off of each other trying to establish dynamic equilibrium. At this point, molecules will continue bouncing off of each other but will remain in equilibrium.

PAINLESS TIP

Imagine that you are in a big, open room with the doors and windows closed. You and your friend are in two opposite corners of the room. You spray from a container of strong cologne. It will smell really strong, at first, in your corner. Your friend won't smell it. Over the next couple of minutes, the smell will lessen in your corner and your friend will begin to smell it. What is happening is that the fragrance molecules are diffusing from the area where they are highly concentrated toward the area with a lower concentration in order to achieve a balanced concentration across the room, thus achieving dynamic equilibrium.

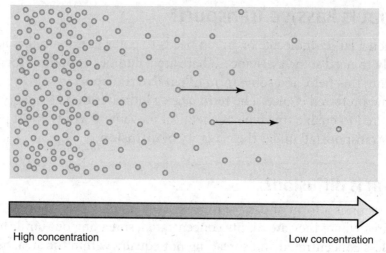

High concentration   Low concentration

Figure 4–4. Diffusion

What is osmosis?

Osmosis is a type of passive transport in which water is able to cross the selectively permeable membrane, while the molecules mixed into the water (this mixture is called a **solution**) cannot cross the same membrane. In osmosis, the water (a **solvent**) moves across the cell membrane in order to create dynamic equilibrium for a dissolved material (a **solute**).

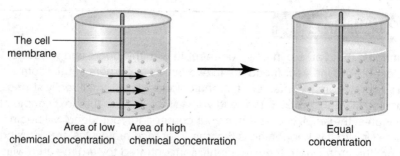

The cell membrane

Area of low   Area of high   Equal
chemical concentration   chemical concentration   concentration

Figure 4–5. Tonicity and Osmosis

As you can see in Figure 4–5, osmosis occurs when it's difficult for large molecules like glucose or ions like salt (in a solution with water) to cross a cell membrane. The water will cross the membrane instead in order to create dynamic equilibrium for the concentrations of the molecules on both sides.

PAINLESS TIP

Isn't it interesting to see that the water level on either side of the cell membrane can be so different? Cells depend on this ability of water to cross the cell membrane to maintain homeostasis. Remember that term? It means keeping the cell balanced and able to survive in spite of whatever is happening outside the cell (the cell's external conditions).

What is the difference between osmotic solutions: hypotonic, isotonic, and hypertonic?

So, when the dissolved ions and large molecules are already in equilibrium on either side of the cell membrane, water doesn't need to create the equilibrium and the solution surrounding the cell is called **isotonic**. The cell in this solution is in a state of **equilibrium**.

When the solution containing a cell has fewer large molecules (like sugar), ionic compounds (like salts), and ions (like sodium and chloride) than the inside of the cell, water will move from the solution into the cell. This makes the cell swell, possibly even bursting the cell. The solution, in this case, is called **hypotonic**.

When the solution containing a cell has more large molecules (like sugar) and ionic compounds (like salt) than the inside of the cell, water will move out of the cell and into the solution. This makes the cell shrink, possibly even imploding the cell. The solution, in this case, is called **hypertonic**.

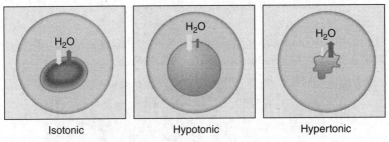

Isotonic   Hypotonic   Hypertonic

Figure 4–6. Facilitated Diffusion

What is facilitated diffusion?

Another type of passive transport is facilitated diffusion. Diffusion is still happening, moving molecules from higher toward lower concentrations. The problem is that the molecules are too large or too charged to squeeze through and diffuse. The process of facilitated diffusion requires some mechanical assistance to help move these large molecules and ionic compounds along. Facilitated diffusion still does not use the cell's energy; instead, it uses structural proteins embedded in the cell membrane. These **transport proteins** may be channel proteins that cross the entire bilayer and act as tunnels. They may also act as carrier proteins recognizing, receiving, and moving a specific particle through the membrane.

FACILITATED DIFFUSION

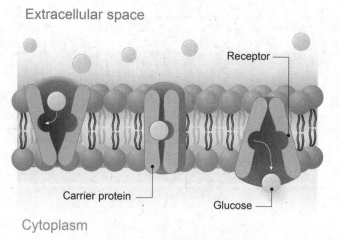

Figure 4–7. Endocytosis and Exocytosis

BRAIN TICKLERS   Set # 2

Decide whether each of the following statements is true or false.

1. A cell may explode because it's in a hypertonic solution.

2. Molecules move from a low to a high concentration in passive transport.

3. Cellular energy is not required for facilitated diffusion.

(Answers are on page 67.)

What Is Active Transport?

Unlike passive transport, **active transport** requires the use of cellular energy to happen. Active transport moves ions and large molecules against the **concentration gradient**, or from a low concentration to a higher concentration. So, the difficulty in transport is both size and/or charge and the concentration gradient. Without assistance from cellular energy, these particles would never be able to cross the cell membrane. The energy used by cells to drive active transport is provided by hydrolysis of adenosine triphosphate (remember ATP?). And then, often the energy that drives active transport is assisted by enzymes. Remember, enzymes reduce the energy needed to start and maintain a reaction.

Why do molecules need to travel against the concentration gradient?

Cells often need large molecules, ionic compounds, and ions even if they already have a higher concentration of those particles than the surrounding fluid. In the human body, an example would be calcium. Calcium ions (Ca++) trigger muscular contractions and are also needed for the structure of bones. Many of the mechanisms that deposit calcium into bones are based on active transport against the concentration gradient.

How is active transport used to move large or charged particles into a cell?

When energy is used to move large molecules and ions into a cell, it is called **endocytosis**. Endocytosis involves the cell membrane bulging inward, into the cell, using the cell's energy to wrap around the material being brought into the cell. Ultimately, the cell membrane entirely engulfs this material, and it is moved further into the cell where it is needed and will be used. Endocytosis is composed of two subgroups—**phagocytosis** and **pinocytosis**. Phagocytosis is the term for engulfing solid particles to bring into the cell. Pinocytosis is the term for engulfing liquid droplets to be used by the cell.

How is active transport used to move large or charged particles out of a cell?

When energy is used to move large molecules and ions out of a cell, it is called **exocytosis**. Exocytosis is the reverse of endocytosis. Any large molecules, ions, and ionic compounds that are destined to be sent out to work in other cells (like hormones), waste particles, or those not needed by the cell are packaged and surrounded by the cell membrane inside the cell, forming a vesicle. The vesicle moves to and fuses with the cell membrane. As this fusion occurs, the vesicle contents are expelled from the cell into the extracellular fluid.

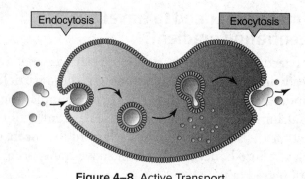

Endocytosis   Exocytosis

Figure 4–8. Active Transport

BRAIN TICKLERS   Set # 3

Name the term that represents each of the following descriptions.

1. Active transport of materials out of the cell

2. Source of energy that drives cellular reactions

3. Active transport of liquid droplets into cells

4. Osmotic balance between the inside and outside of a cell

5. When a cell engulfs a solid particle during active transport

A. phagocytosis

B. isotonic

C. ATP

D. exocytosis

E. pinocytosis

(Answers are on page 67.)

What Happens When Cell Membranes Don't Work Correctly?

Functional cell membranes are imperative to every cell's ability to maintain homeostasis and stay alive. Cell membrane damage may be the result of external environmental issues or a genetic condition. Osmotic changes can alter cell membrane function, as can certain chemical agents, viral proteins, and bacterial toxins. Organelle dysfunction can also affect the membrane. When the cytoskeleton or mitochondria aren't working, they can cause the cell membrane to stop transporting effectively. In humans, cystic fibrosis and Wilson's disease are two genetic conditions affecting the cell membrane. In these diseases, malfunctioning transport channel proteins disrupt the maintenance of intracellular homeostasis.

SUPER BRAIN TICKLERS

Match these definitions to the correct terms.

1. The density change of particles between one area and another

2. When the particle density is in balance throughout a defined area

A. diffusion

B. solvent

C. endocytosis

3. When material is engulfed within a vesicle of the cell membrane and then transported into the cell

4. Movement of molecules or particles from an area of high concentration to an area of low concentration

5. Water-loving; able to interact with water

6. A type of transport across cell membranes that requires the use of energy

7. Inside the cell

8. The liquid in which solutes are dissolved to form a solution

D. active transport

E. concentration gradient

F. intracellular

G. hydrophilic

H. dynamic equilibrium

(Answers are on page 67.)

Vocabulary

Active transport: A type of transport that requires energy; it moves ions and large molecules against the concentration gradient, or from a low concentration to a higher concentration.

Concentration gradient: The difference in the concentration of a substance between two areas.

Diffusion: A form of passive transport where molecules move from an area where they are highly concentrated, spreading out until they are less concentrated and spreading out equally within the area.

Dynamic equilibrium: When molecules are evenly distributed throughout an area by diffusion.

Endocytosis: A form of active transport in which materials are moving into the cell.

Equilibrium: In biological transport, a stable state in which all influences are balanced by equally opposing forces. The net movement into, around, and out of the system is equal and balanced.

Exocytosis: A form of active transport in which materials are moving out of the cell.

Fluid mosaic model: The name by which scientists call their representation of the plasma membrane; all of the components of a cell membrane are able to stay in constant "fluid" motion, and the parts that make up the outside layers of the cell membrane look like an artistic mosaic.

Hydrophilic: Molecules that have an imbalanced, or partial, charge that allows them to attract water molecules.

Hydrophobic: Molecules that have a polarity that repels water.

Hypertonic: When the solution containing a cell has more large molecules and ionic compounds than the inside of the cell, water will move out of the cell and into the solution. This makes the cell shrink, possibly even imploding the cell.

Hypotonic: When the solution containing a cell has fewer large molecules, ionic compounds, and ions than the inside of the cell, water will move from the solution into the cell. This makes the cell swell, possibly even bursting the cell.

Intracellular transport: Movement of materials inside a cell.

Intercellular transport: Movement of materials between different, often adjoining cells.

Isotonic: When the dissolved ions and large molecules are already in equilibrium on either side of the cell membrane, water doesn't need to create the equilibrium and the cell should neither gain nor lose mass.

Osmosis: A type of passive transport in which water is allowed to cross the selectively permeable membrane while the molecules mixed into the solution are not permitted to cross the same membrane.

Passive transport: A type of diffusion in which no energy is required to enable chemical movement.

Phagocytosis: The cellular process of engulfing solid particles to bring into the cell.

Pinocytosis: The cellular process of engulfing liquid droplets to be used by the cell.

Selective permeability: The ability of a cell membrane to determine which molecules are able to cross over from one side of the cell membrane to the other.

Solute: A substance that is being dissolved.

Solution: Formed when one substance dissolves into another.

Solvent: The substance in which a solute dissolves to form a solution.

Transport proteins: Structures involved in facilitated diffusion, not requiring energy to move molecules across the cell membrane. They may be channel proteins that cross the entire bilayer and act as tunnels. They may also act as carrier proteins recognizing, receiving, and moving a specific particle through the membrane.

Brain Ticklers—The Answers
Set # 1, page 56
1. B
2. C
3. A

Set # 2, page 61
1. False
2. False
3. True

Set # 3, page 63
1. D
2. C
3. E
4. B
5. A

Super Brain Ticklers
1. E
2. H
3. C
4. A
5. G
6. D
7. F
8. B

Chapter 5

Metabolism and Energy

The Relationship Between Metabolism and Energy

Why is energy so important? you may ask. Well, energy is what allows all work to happen. Mechanically, many of us use gasoline to provide the energy to drive cars and electricity provided by power plants to use our computers and phones and to light, heat, and cool our homes. Our cells and bodies also need energy to do the jobs they need to perform. Remember learning about metabolism in Chapter 1? It is one of the characteristics that all living things share. Metabolism is the ability to take in nutrients, to change those nutrients into a form we can use for energy, to use that energy, and then to discard any remaining waste.

PAINLESS FACT

Imagine a marathon runner. These athletes spend months preparing for their 26.2-mile run by nourishing their bodies well, working out, and running longer and longer distances. Finally, it's the night before the race and the runners all eat huge pasta dinners. This is called carb loading. The runners are taking in complex carbohydrates that will take hours to slowly break down, providing them with small, usable amounts of energy to help them get through the race. During the race, the marathoners will take in small amounts of simple sugars to give themselves useful amounts of quickly metabolized energy. Marathon runners have figured out how to use the energy from nutrients to best enhance their athletic performance.

Maybe you are into fitness or perhaps you simply want to consider ways to stay better focused in school or during an evening of studying. Activities such as sitting up, blinking and focusing your eyes, turning pages, and holding a textbook all require the use of energy. When we talk about living things using energy, we are referring to how their cells obtain, metabolize, and use energy.

Energy

Cells can't just run out to pick up a candy bar or quick spaghetti dinner in order to get some usable energy. They need a way to obtain nutrients and energy quickly and in a form that's easily utilized and managed. Most living things use carbohydrates as their primary energy source. Carbohydrates are made naturally by **producers**, or **autotrophs**, during **photosynthesis** and are used by these autotrophs and by **consumers**, or **heterotrophs**, during **cellular respiration**.

BRAIN TICKLERS   Set # 1

Name the term that represents each of the following descriptions.

1. The process used by all living things to gain the energy needed to stay alive

2. Can make their own food by using energy from sunlight to conduct photosynthesis

3. An organism that cannot produce its own food

4. Capturing light energy from the sun and using it to synthesize glucose and release oxygen

A. producer

B. heterotroph

C. photosynthesis

D. cellular respiration

(Answers are on page 85.)

Energy exchange within the environment

The simple photosynthesis/cellular respiration cycle is the basis of energy exchange for almost every life form on Earth. The process of photosynthesis involves energy from the sun interacting with the water and carbon dioxide present in the chlorophyll inside the leaves of plants. The carbon, hydrogen, and oxygen present in water (H_2O) and carbon dioxide (CO_2) are rearranged to produce simple sugar ($C_6H_{12}O_6$) and leftover oxygen (O_2). Sugar and oxygen are then used to provide energy for cellular respiration to occur. Leftover molecules from cellular respiration are water and carbon dioxide, which are recycled to become the molecules used by autotrophs in the next round of photosynthesis.

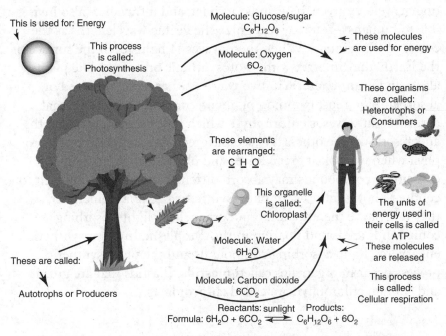

Figure 5–1. Basic Overview of Photosynthesis/Cellular Respiration Cycle

Photosynthesis occurs in plants, algae, and many types of bacteria. These autotrophs capture energy from sunlight, carbon dioxide from the air, and water mostly from the soil. Inside the plant cell, in the chloroplast, the CO_2 is reduced (meaning that it gains electrons) and the H_2O is oxidized (meaning that it loses electrons). The process of photosynthesis occurs in three steps: the first two are **light-dependent reactions**, which are chemical energy-storing phases that occur only when sunlight is available, and the third step, the **light-independent reactions**, happen whether or not the sun is shining. Since the light-independent reactions use the products of the light-dependent reactions, sunlight does impact all three steps. All three processes are involved in the production of carbohydrates, ultimately generating glucose and oxygen. Most of the oxygen is released into the atmosphere and the glucose stores energy to help drive the cellular respiration processes called **glycolysis**, the **Krebs cycle**, and **oxidative phosphorylation**.

Photosynthesis

The sun's energy reaches the surface of the Earth in the form of wavelengths of light: infrared (wavelengths invisible to humans), red, orange, yellow, green, blue, indigo, violet, and ultraviolet (also invisible to humans). You may recognize the visible wavelengths as the colors of the rainbow. When these waves of light reach the surface of the Earth, they interact with photosynthetic organisms like plants, algae, and **cyanobacteria** in two ways—either being reflected or absorbed. The most common pigment-containing molecule that absorbs light waves is **chlorophyll**, which reflects green wavelengths and absorbs all the others. We see the green light reflected by chlorophyll when we look at green leaves and plants. Other pigment-containing compounds may absorb different wavelengths of light to conduct photosynthesis. In photosynthetic bacteria (remember, they don't have membrane-bound organelles), light-absorbing compounds are found floating in the cytoplasm. In photosynthetic eukaryotes, light-absorbing compounds and the photosynthetic process occur in organelles called **plastids**. Plastids that are green and contain chlorophyll are called **chloroplasts**.

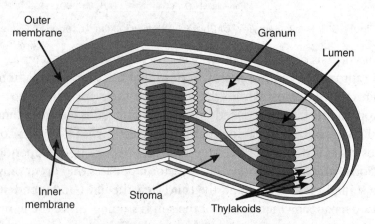

Figure 5–2. Structure of a Chloroplast

BRAIN TICKLERS   Set # 2

Name the term that represents each of the following descriptions.

1. The first step of both aerobic and anaerobic respiration; this step splits glucose into pyruvate

2. The compound in green plants that traps energy from sunlight to fuel photosynthesis

3. First two stages of photosynthesis

4. The organelle in plants that contains pigment-containing compounds and where photosynthesis occurs

A. chloroplast

B. glycolysis

C. light-dependent reaction

D. chlorophyll

(Answers are on page 85.)

Light-dependent reactions

When chloroplasts absorb light waves from the sun, the light energy is stored temporarily in structures called **thylakoid** membranes. Layers of thylakoid membranes look like stacked pancakes and are called **grana**. Light-dependent reactions are initiated by the energy stored in the thylakoids. This first stage of photosynthesis begins with **photolysis**, when light energy splits a water molecule into individual hydrogen and oxygen molecules, releasing an electron in the chloroplast and an oxygen molecule into the atmosphere.

Now, light energy is no longer necessary to drive the reaction. Chemical energy from the remaining electron is responsible for generating the second stage of light-dependent photosynthetic activity. This stage is called the **electron transport chain**. Here, the loose electron from photolysis is passed between a series of electron carriers, losing a tiny bit of energy each time it moves to another electron carrier. The energy that is lost is then used to trigger two energy-storing activities, creating ATP and building NADPH. You might remember ATP from Chapter 3 as a molecule that stores cellular energy. The leftover electron from photolysis provides the energy to add the third phosphate group to ADP to form ATP.

Similarly, the other energy-storing activity is adding a hydrogen ion to a molecule of NADP to form NADPH.

Light-independent reactions

Light-independent reactions (previously called dark reactions) include carbon-fixing reactions and the Calvin cycle, and they occur whether or not sunlight is available. These reactions occur in the **stroma**, the cytoplasmic fluid of the chloroplast that surrounds the grana. Light-independent reactions use the energy stored in the ATP and NADPH synthesized during the second stage of the light-dependent reactions to rearrange carbon dioxide and hydrogen into a series of carbohydrate molecules, ultimately producing glucose. Energy used to form the glucose comes from the hydrolysis of ATP and oxidation of NADH to form ADP and NADP, respectively.

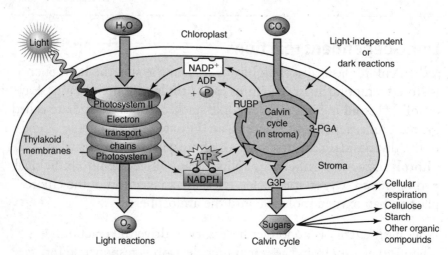

Figure 5–3. Summary of the Major Steps in Photosynthesis

C3 and C4 photosynthesis

Most plants conduct photosynthesis by producing an intermediate three-carbon carbohydrate (3-PGA) during the light-independent reactions to ultimately yield glucose. The full name of the three-carbon carbohydrate is 3-phosphoglyceric acid. Plants that make glucose with just a 3-PGA intermediate sugar use C3 photosynthesis. Some species of plants that live in areas with reduced water, intense light, high temperature, or high salinity (salt levels) may have difficulty accessing enough water or carbon dioxide to conduct C3 photosynthesis. These plants, including corn and sugarcane as well as many cactuses and grasses, use C4 photosynthesis instead of C3. During C4 photosynthesis, a four-carbon intermediate compound and carbon dioxide are produced during the light-independent stage. While it is less efficient under normal conditions, C4 photosynthesis keeps more carbon available to these plants so that they can still manufacture sugar and stay alive in extreme environments.

PAINLESS STUDY TIP

The formula for photosynthesis is:

$$6CO_2 + 6H_2O \rightarrow C_6H_{12}O_6 + 6O_2$$

The formula for cellular respiration is:

$$C_6H_{12}O_6 + 6O_2 \rightarrow 6CO_2 + 6H_2O$$

Do you see the relationship between the two formulas? That's right! They are the same formula but reversed!

Chemosynthesis

Most ecosystems in the world are based on a food chain that obtains its energy from sunlight in order to conduct photosynthesis. How, then, does life exist where sunlight can't reach? Chemosynthetic autotrophs live in places like deep ocean vents and seafloors, hot springs, and areas deep in the soil, in swamps, in the guts of cattle, and even in the human digestive system! Photosynthesis and chemosynthesis, together, provide the energy for all life on Earth. Chemosynthesis is the process during which specific microbes create carbohydrates as a result of inorganic chemical reactions. Different chemoautotrophs use unique chemical pathways to create sugar; some use hydrogen sulfide, some use methane, some use iron, and others use inorganic nitrogen.

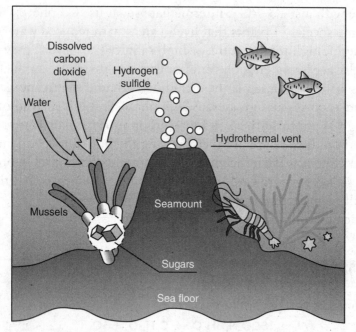

Figure 5–4. The Process of Chemosynthesis

BRAIN TICKLERS   Set # 3

Decide whether each of the following statements is true or false.

1. The formula for cellular respiration is simply the opposite of the formula for photosynthesis.

2. Light-independent reactions only happen in darkness.

3. Energy from sunlight is the only energy available to autotrophs.

(Answers are on page 85.)

Cellular Respiration

So, we've established that all living things get their energy from photosynthesis and chemosynthesis. The next question you're probably going to ask would be, How do all these organisms store and use that energy? Or perhaps you're wondering, What does this have to do with metabolism? Well, the short answer is that by undergoing cellular respiration using the products of photosynthesis and chemosynthesis, all living things produce adenosine triphosphate, or ATP. In living bodies, the molecule ATP acts like the change we put into a vending machine to get a small snack. Each time that the bond holding the third phosphate of ATP is broken, a small amount of energy is released, giving the cell just enough energy to do the required work. **Aerobic respiration** is the three-step process that breaks down sugar in the presence of oxygen to give us ATP. Cellular respiration that occurs in the absence of oxygen is called **anaerobic respiration**.

Aerobic respiration

Glycolysis is the first of three steps of aerobic respiration.

 **REMINDER**

Remember you learned what *-lysis* meant in Chapter 3? So, what would glycolysis mean? Right! It means to break apart a glucose molecule!

During glycolysis, a six-carbon glucose molecule (that was produced during photosynthesis) is split into two three-carbon molecules called pyruvates. This step produces four molecules of ATP, uses two molecules of ATP, and occurs in the cell's cytoplasm. The final products of glycolysis are two available high-energy molecules each of ATP and NADH.

The **Krebs cycle** is the second step of aerobic respiration. In this step, pyruvates from glycolysis are moved into the mitochondria. Here, a carbon dioxide molecule is removed and the three-carbon pyruvate becomes a two-carbon molecule called acetyl-CoA. The final products of the Krebs cycle are two molecules of ATP, $FADH_2$ (which yields two molecules of ATP), and NADH (which initiates the production of eight molecules of ATP).

The third and final step in breaking down sugar to become energy for cells to use is called **oxidative phosphorylation**. This step occurs as the electron transport chain, embedded in the inner mitochondrial membrane, generates energy by reducing NADH to NAD+ and

FADH$_2$ to FAD+. Each time energy is released in these reactions, it is recaptured and used to make ATP. Each reduction of NADH contributes three ATP molecules, and each molecule of FADH$_2$ adds two molecules of ATP. The process of aerobic respiration produces 40 molecules of ATP but since four ATPs are used up to drive the reactions, the final yield is 36 available molecules of ATP. A byproduct released in this process is water.

PAINLESS STUDY TIP

Remember this? It's really important!

The formula for cellular respiration is:

$$C_6H_{12}O_6 + 6O_2 \rightarrow 6CO_2 + 6H_2O$$

The formula for photosynthesis is:

$$6CO_2 + 6H_2O \rightarrow C_6H_{12}O_6 + 6O_2$$

Do you see how the molecules in each reaction are constantly being recycled?

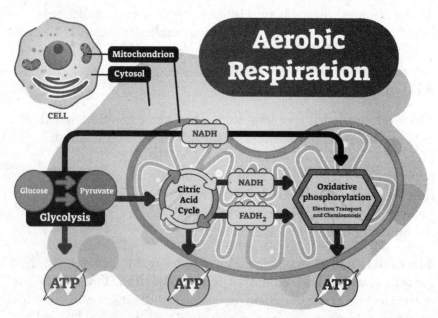

Figure 5–5. Three Steps of Aerobic Respiration

BRAIN TICKLERS   Set # 4

Name the term that represents each of the following descriptions.

1. Cellular respiration in the absence of oxygen

2. A byproduct of photosynthesis and a reactant in cellular respiration

3. Cellular respiration in the presence of oxygen

4. Specific microbes create carbohydrates as a result of inorganic chemical reactions

5. Some species of plants that live in areas with reduced water, intense light, high temperature, or high salinity (salt levels) use this less efficient method of photosynthesis

A. aerobic respiration

B. anaerobic respiration

C. chemosynthesis

D. C4 photosynthesis

E. water

(Answers are on page 85.)

Anaerobic respiration

When oxygen isn't available to help drive cellular respiration, organisms use anaerobic respiration. Although there are a number of anaerobic pathways, none of them are nearly as efficient as aerobic respiration. Many microbes live in oxygen-free environments and some use molecules other than sugar to move electrons and obtain their ATP. Some of the other pathways include denitrification and sulfate, carbonate, and nitrogen reduction reactions. Usually, though, when we talk about anaerobic respiration we are referring to the use of sugars to obtain energy, just without using oxygen to help move the electrons. The two kinds of carbohydrate-based anaerobic respiration are both forms of fermentation; both happen in the cytoplasm and both begin with glycolysis. Both **alcoholic fermentation** and **lactic acid fermentation** produce four molecules of ATP—two are used to drive the reactions and two become available to the cell as stored energy.

What Are the Two Types of Anaerobic Respiration?

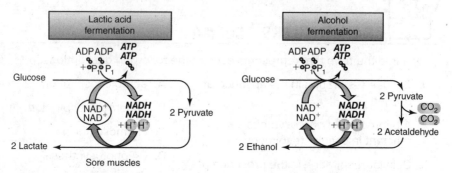

Figure 5–6. Two Common Types of Anaerobic Respiration

Alcoholic fermentation occurs in yeast to produce bread, and many grains and fruit produce alcoholic beverages when they are fermented. Alcoholic fermentation begins with glycolysis (remember the splitting of glucose?) and produces ethanol, carbon dioxide, and ATP. In humans, anaerobic respiration comes in the form of lactic acid fermentation. This also begins with glycolysis and yields lactic acid and ATP. The most common reason for a human to begin needing anaerobic respiration is an extreme workout. When gasping for breath isn't enough and we lack the oxygen for our cells to conduct aerobic respiration, our cells have the anaerobic backup plan. Unfortunately, when we accumulate lactic acid as a result of this process, the lactic acid collects between our muscle fibers and can cause quite a bit of pain.

PAINLESS FACT

Have you ever exercised and thought you couldn't go on, but you did keep on going? Were you sore for the next couple days? Now you know why!

Energy-storing Macromolecules

Finally, let's revisit ATP one more time because it is such an important molecule to all living things on Earth.

ADP/ATP

Adenosine diphosphate (ADP) is a relatively stable molecule made up of an adenosine molecule bonded to a ribose with two sequential phosphate groups attached. Energy must be used to attach a third phosphate group to the line to form ATP. When the third phosphate bond is broken to form ADP+ inorganic phosphate, energy is released for immediate use by the cell.

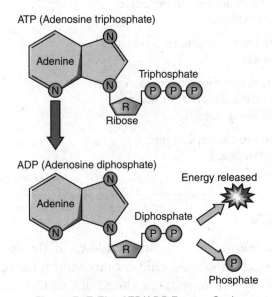

ATP (Adenosine triphosphate)

Adenine

Triphosphate

Ribose

ADP (Adenosine diphosphate)

Adenine

Diphosphate

Energy released

Phosphate

Figure 5–7. The ATP/ADP Energy Cycle

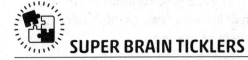

SUPER BRAIN TICKLERS

Match these definitions to the correct terms.

1. Organisms that cannot make their own food and are entirely dependent on photosynthetic organisms for obtaining carbohydrates to give them energy

2. Anaerobic respiration in humans that provides a backup plan for energy if they run low on available oxygen

A. photolysis

B. carbohydrate

C. consumer

3. The biomolecule that gives consumers quick or stored energy

4. Internal structures of a chloroplast; the site of sunlight energy storage

5. The stage of photosynthesis when light energy splits a water molecule into individual hydrogen and oxygen molecules, releasing an electron in the chloroplast and an oxygen molecule into the atmosphere

6. Prokaryotes that are able to conduct photosynthesis

7. A relatively stable molecule made up of an adenosine molecule bonded to two phosphate groups

8. The fluid in the chloroplast that surrounds the grana

D. stroma

E. cyanobacteria

F. thylakoid membrane

G. ADP

H. lactic acid fermentation

(Answers are on pages 85–86.)

Vocabulary

Aerobic respiration: The cellular breakdown of glucose in the presence of oxygen. It is the most efficient mechanism for getting energy in the form of ATP from sugar produced during photosynthesis, ultimately yielding 36 usable molecules of ATP.

Alcoholic fermentation: A form of anaerobic respiration (in yeast and in some fruit) that gets a small amount (two usable molecules) of ATP from sugar without the help of oxygen.

Anaerobic respiration: Cellular respiration in the absence of oxygen; it only produces two usable molecules of ATP. The most common forms are lactic acid fermentation and alcoholic fermentation.

Autotroph: An organism that makes its own food; most autotrophs are photosynthetic while some are chemosynthetic.

Cellular respiration: The process used by all living things to gain the energy needed to stay alive. Cycles continually with photosynthesis.

Chlorophyll: The pigment-containing compound found in green plants. It traps energy from sunlight to fuel photosynthesis.

Chloroplast: The organelle in plants that contains chlorophyll and houses the process of photosynthesis.

Consumers: Also known as heterotrophs, these organisms cannot make their own food and are entirely dependent on photosynthetic organisms for obtaining carbohydrates to give them energy.

Cyanobacteria: Prokaryotes that are able to conduct photosynthesis.

Electron transport chain: The second stage of light-dependent photosynthetic activity, when loose electrons from photolysis are passed between a series of electron carriers, losing a tiny bit of energy each time they move to another electron carrier.

Glycolysis: The first step of both aerobic and anaerobic respiration; this step splits glucose (one molecule containing six carbons) into pyruvate (two three-carbon molecules).

Granum (grana pl.): The site of light-dependent reactions in photosynthesis; grana are the stacked thylakoid membranes inside the chloroplast.

Heterotroph: An organism that cannot produce its own food. Also known as a consumer.

Krebs cycle: The second step of photosynthesis, this light-dependent reaction splits 3-carbon pyruvate into 2-carbon acetyl-CoA molecules, releasing CO_2 and generating energy via ATP, NADH, and $FADH_2$.

Lactic acid fermentation: Anaerobic respiration in humans that provides a backup plan for energy if they run low on available oxygen. Produces small amounts of energy, and lactic acid is a waste product.

Light-dependent reactions: First two stages of photosynthesis that require sunlight.

Light-independent reactions: Final stage of photosynthesis that can occur with or without sunlight.

Oxidative phosphorylation: Final stage of aerobic respiration; here the electron transport chain moves available energy into the third phosphate bond of ATP.

Photolysis: A stage of the light-dependent reactions of photosynthesis when light energy splits a water molecule into individual hydrogen and oxygen molecules, releasing an electron in the chloroplast and an oxygen molecule into the atmosphere.

Photosynthesis: The action of producers in capturing light energy from the sun and using it to synthesize glucose and release oxygen.

Plastids: Organelles in plants that contain light-absorbing compounds, allowing photosynthesis to occur inside them.

Producers: Autotrophs that can make their own food by using energy from sunlight to conduct photosynthesis.

Stroma: The fluid in the chloroplast that surrounds the grana; they are the sites of the light-independent reactions.

Thylakoids: Internal membranes of a chloroplast; structures that are the sites of storage of energy from sunlight.

Brain Ticklers—The Answers

Set # 1, page 70

1. D
2. A
3. B
4. C

Set # 2, page 73

1. B
2. D
3. C
4. A

Set # 3, page 76

1. True
2. False
3. False

Set # 4, page 79

1. B
2. E
3. A
4. C
5. D

Super Brain Ticklers

1. C
2. H
3. B

4. F
5. A
6. E
7. G
8. D

Chapter 6

Human Anatomy

The Human Body

Understanding how the bodies of humans are assembled and how they work is an important element of biology because, remember, biology is the study of all living things and how they function.

 REMINDER

Do you remember from Chapter 1 that organization is an important characteristic of all living things? And how that organization involves similar cells grouped into tissues, tissues working together to form organs, organs working together to create systems, and systems working together to form the organism? Well, the human body is an interesting and important example of how those levels of organization interact.

Most anatomists recognize the human body as being made up of 11 separate body systems. Each of those systems performs different functions, each system is made up of different kinds of cells, and each can cause different kinds of diseases and disorders when it is not functioning correctly.

Basic Anatomy and Physiology

Anatomy and physiology are the studies of how living things work. Anatomy explores the structures and parts of living things, while physiology studies how those structures work. Before we investigate each of the unique human body systems, we need to pay attention to some aspects of anatomy and physiology that these systems have in common.

Planes and Directional Terms

When we talk about the human body, it's important to use a common language that clearly describes our references. Planes are dissectional terms that refer to the way that an organ or body is "sliced" in an imaging study or a dissection. A midsagittal plane separates the body into right and left halves, and as the cut moves away from the body's midline it's simply called a **sagittal** plane. A **transverse**, or horizontal, plane separates the body into **superior** (toward the head) and **inferior** (toward the feet) sections. A **frontal**, or coronal, plane divides the front (**anterior**) from the back (**posterior**) sides of the body. An **oblique** plane is on an angle.

Directional terms are generally paired words that indicate a direction in the body. **Medial** refers to going toward the midsagittal line while **lateral** refers to moving away from the midsagittal line.

For the arms and legs, **proximal** means toward the torso (the trunk of the body) while its paired term, **distal**, means moving toward the hands or feet. Many directional terms are named for the skeletal structures nearby.

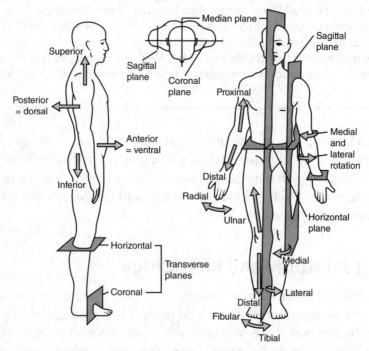

Figure 6–1. Directional Terms and Planes

Body Cavities and Membranes

The human body contains several cavities. The **dorsal cavity** contains the cranial cavity (brain) and the vertebral cavity (spinal cord). The **ventral cavity** is found in the torso, or trunk. The ventral cavity is divided in half by the muscular diaphragm. Above the diaphragm, the chest contains the pleural (lung), mediastinal (superior to the heart), and pericardial (heart) cavities. Inferior to the diaphragm lie the abdominal cavity (with the stomach, small intestine, most of the large intestines, liver, gallbladder, and spleen), the pelvic cavity (distal ureters, urinary bladder, urethra, distal colon, and internal reproductive organs), and the retroperitoneal cavity (kidneys, adrenal glands, and most of the pancreas).

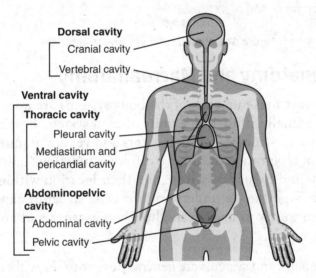

Figure 6–2. Human Body Cavities

The body's membranes are interesting. **Serous membranes** are smooth tissue linings that secrete lubricating fluid. Their job is to insulate, protect, maintain position, and lubricate all of our organs. Serous membranes that line the insides of the body cavities are called **parietal membranes**. Serous membranes that surround the organs in the body are called **visceral membranes**. Because parietal and visceral membranes are continuous with each other, organs are kept in place and the nerves, lymphatic vessels, and blood vessels that serve them are also held in position.

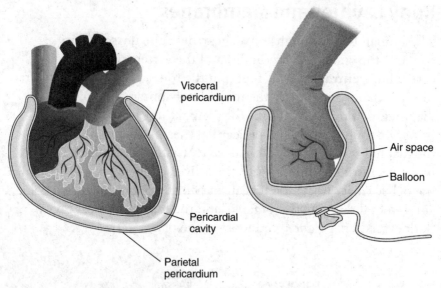

Figure 6–3. Parietal and Visceral Membranes

Microanatomy and Macroanatomy

The cells that make up each of the body systems are microscopic. These, the smallest functional units of each system, have similar organelles and structural arrangements and yet are unique from one system to another. Cells have different functions based on the organs and tissues they form and their location within the body. It is important to understand the cells' structures and functions when we try to understand how bodies work and how diseases happen.

Tissues, organs, and systems are generally easier to see without a microscope. These structures are called macroscopic, or **gross anatomy**. The word *gross* here means large enough to be seen with the naked eye.

BRAIN TICKLERS　Set # 1

Name the term that represents each of the following descriptions.

1. Serous tissue that surrounds the organs

2. The space in the body that contains the brain and spinal cavity

3. A cut through the body that divides the superior from the inferior

A. dorsal cavity

B. transverse plane

C. visceral membrane

(Answers are on page 127.)

Skeletal System

Functions

The skeletal system is made up of bones and the structures that hold them together. Normally, adults have 206 bones in their bodies, less than the approximately 300 bones in a young child. The reason adults have fewer bones is that some of the bones in a child fuse together as they mature. Bones are living structures that serve several functions; they support the body and allow movement, and they protect the internal organs. Blood cells are produced in the long bones and bones store fat and calcium. Joints are the places where the bones are held together and where bending and movement originate.

PAINLESS FACT

Many people are not aware that our bones and skeleton are living tissue. They are! Their cells perform all the same functions as any other living cell.

Microscopic

The bones that make up the skeleton need to be very strong and tough. This strength is achieved because the structural cells of most of the skeleton are arranged in cylindrical osteons, also known as haversian systems. Each osteon is arranged with 5–20 concentric rings of calcified lamella surrounding an inner channel where blood vessels and nerves serve the living bone cells. New bone is deposited by cells called osteoblasts. Mature bone cells are called osteocytes. The visceral membrane surrounding the bones is called the periosteum. During our human lifetime our skeletons are continually modifying themselves to respond to our environment, to how our body is moving, and even to the shoes we wear. Cells called osteoclasts destroy bone where it's not needed, and new osteoblasts build new bone where it is required. This continuous modeling process results in totally rebuilt skeletons in humans approximately every 10 years.

PAINLESS STUDY TIP

Have you noticed that many of the scientific words referring to the skeleton contain the root "-*osteo*"? *Osteo* is a term that tells you that you are referring to bones!

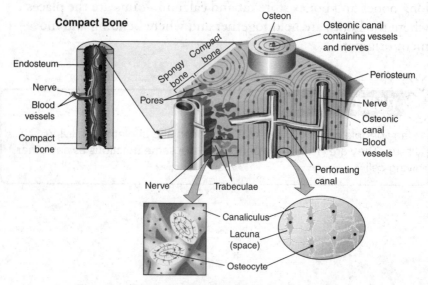

Figure 6–4. Microscopic Anatomy of Compact Bone

Macroscopic

The human skeleton is divided into two regions: the axial skeleton and the appendicular skeleton. The axial skeleton includes the core bones of the body: all bones in the skull, hyoid, sternum, vertebrae and ribs. The appendicular skeleton consists of the bones in the arms and legs (appendages) as well as the bones that attach these appendages to the axial skeleton. Bones come in a number of shapes: long bones such as the femur and radius, short bones like those in the toes, flat bones like those found in the skull, irregular bones like the vertebrae, and sesamoid bones such as the patella (kneecap). Inside the long bones are spaces filled with yellow and red bone marrow. Yellow marrow stores fat reserves for the body. Red bone marrow is the site where stem cells differentiate into red and white blood cells and platelets. This process is called **hematopoiesis**.

The connective tissue that holds one bone to its adjoining bone is called a **ligament**. Places where bones come together are called joints, or **articulations**. Some joints are fixed, or unmoving. These synarthrotic joints include the bones of the cranium, which fuse together throughout childhood to become sutures—tough, immobile locations where the bones are attached solidly together. Some articulations are called amphiarthrotic; these partially mobile joints have cartilage between their adjoining surfaces, making them a bit more flexible than synarthrotic joints. One amphiarthrotic joint is the attachment of the ribs to the sternum, or breastbone. Cartilage, a tough and flexible connective tissue, allows this joint to have limited movement and helps with bone growth and repair. Finally, the most common and most movable type of articulation is the diarthrotic joint. Most diarthrotic joints have cartilage and an additional cushion made of a membrane-bound capsule called a bursa filled with a cushioning lubricant called synovial fluid. Some types of diarthrotic joints include the ball and socket at the hip and shoulder, the hinge joints at the knees and elbows, pivot joints at the base of the skull and the top vertebrae, and gliding joints of flat surfaces moving across each other, like one vertebra against the next.

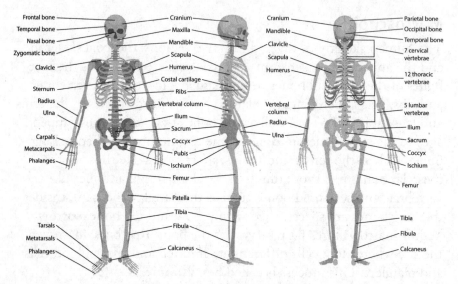

Figure 6–5. The Human Skeleton

Disease

Skeletal diseases and disorders may affect the bones, the joints, or both. Injuries include broken bones, dislocations, and sprains. One common sports injury is ACL damage, which is an injury to the anterior cruciate ligament behind the knee. As in any body system, problems can arise from injury, genetic and birth defects, infection, or cancer, or they may be **idiopathic** (meaning no known cause). Here are a few examples of skeletal disorders:

- **Osteoporosis**—A porous bone disease that involves bone deterioration and low bone mass. Most often found in women, it can be the result of reduced estrogen after menopause. Osteoporotic bones are thinner and more easily fractured. Treatment with calcium, vitamin D, and bisphosphonates, as well as weight-bearing exercise, can be helpful in rebuilding bone strength and mass.

- **Osteoarthritis**—A degenerative joint disease associated with aging and past joint trauma. Cartilage between the bones deteriorates and bony surfaces form jagged spurs, causing both swelling and pain. While osteoarthritis has no cure at present, antiinflammatory painkillers may help and exercise is usually recommended.

- **Osteogenesis imperfecta**—Also known as brittle bone disease, OI is an incurable genetic disorder. There are several inheritance patterns causing four different types and severities. Babies with OI are often born with soft, malformed, or broken bones, and broken bones occur easily and frequently throughout life. People with OI often have a blue tinge to the whites of their eyes. Treatment includes setting, splinting, and bracing bones to prevent deformity and ease pain. Medication, as well as physical and occupational therapy, may help in reducing the impact of this disorder.

Muscular System

Functions

The muscular system works to help the skeleton maintain the body's shape and posture and to move the body correctly. It also works independently of the skeleton to move materials (like blood, nutrients, and waste) inside the body. Active muscles generate heat for the body to maintain body temperature and homeostasis.

Microscopic

There are three main types of muscle cells and tissues in the human body: smooth, cardiac, and skeletal muscle. Smooth muscle is made up of small cells, wide in the middle and extending into two pointy ends. They have one nucleus in the middle of each cell. Smooth muscle cells do not have striations (stripes), do not tire easily, can stay contracted for a long time, and are not attached to bones. Smooth muscle cells are not under conscious control, and so they are called involuntary. Smooth muscle makes up much of the internal human musculature: the stomach, large and small intestines, uterus, urinary bladder, gallbladder, and blood vessels. Cardiac muscle is only found in the heart. These muscles are also involuntary, but they are branched tubules and are striated. Cardiac cells have a unique ability to communicate with each other and to create a normal rhythmic heartbeat using special structures called **intercalated discs**. Skeletal muscle is attached to the bones of the skeleton. It is multinucleate, consciously controlled, or voluntary, and striated. Contraction of muscles ordinarily requires aerobic respiration to generate needed energy in the form of ATP. During vigorous exercise, a skeletal muscle may become fatigued because it can't get enough oxygen, a situation called **oxygen debt**. When this happens, the muscle starts using anaerobic respiration, which causes lactic acid to accumulate in the muscle.

PAINLESS FACT

Have you ever had muscle pain for several days after an extreme workout? That's because lactic acid crystals settle in the muscles and interrupt the ordinarily smooth movement of the muscle fibers against each other. The crystals generally dissolve over the next three to four days, with discomfort lessening as crystals become smaller and fewer.

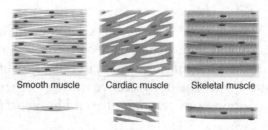

Smooth muscle   Cardiac muscle   Skeletal muscle

Figure 6–6. The Three Types of Muscular Cell and Tissue

Macroscopic

There are more than 650 muscles in the human body. In order to move, most skeletal muscles have two or more points of attachment to bone; the structure that attaches the muscle to a bone is called a *tendon*. Every muscle has a name, usually based on the muscle's size, location, number of origins, locations of origins and insertions, or action it performs. Movement of muscles is triggered by impulses sent by the nervous system. The place where a muscle fiber meets the nervous system is called a **neuromuscular junction**.

Skeletal muscles work in pairs, with each member of the pair moving in an opposite direction when it contracts. An example of this would be that when the biceps brachii (the agonist) on the top and inner side of your arm is contracted, the triceps brachii (the antagonist) relaxes and your arm bends at the elbow. When the opposite happens, the triceps brachii contracts while the biceps brachii relaxes, and the arm straightens. The attachments of skeletal muscles to bone are called either the **origin** or the **insertion**. The origin is the relatively immobile part of the muscle attached to a fixed bone or structure. The insertion is the moving part of the muscle; it is attached to the bone or structure being moved during a muscular contraction.

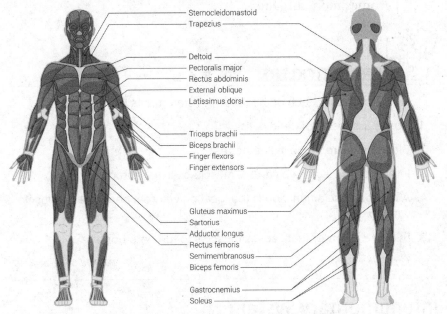

Figure 6–7. Some Major Skeletal Muscles

Disease

Diseases and disorders of the muscular system include injuries like pulls, strains, and tears of muscles and tendons. As in any body system, problems can arise from injury, genetic and birth defects, infection, or cancer, or they may be idiopathic. Here are a few examples of muscular disorders:

- **Tetanus**—An infectious bacterial disease that causes continuous cramping of the skeletal muscles. It is preventable with a vaccination.

- **Duchenne's muscular dystrophy (DMD)**—A sex-linked genetic disorder that causes muscle cells to die. First seen in early childhood, progressive deterioration of muscles weakens the child. Most children with DMD are wheelchair bound by ages 9–11. There is no cure, but pharmaceuticals, physical and respiratory therapies, and orthopedic appliances help in treatment.

- **Fibrodysplasia ossificans progressiva (FOP)**—A very rare genetic condition in which muscle turns to bone, forming a secondary skeleton. There is no cure, and the patient's body eventually freezes into a sitting or a standing/lying down position. As the extra skeleton surrounds the rib cage, breathing becomes more difficult.

BRAIN TICKLERS   Set # 2

Decide whether each of the following statements is true or false.

1. Bones are made of dead cells.

2. Blood cells are made inside bones.

3. Intercalated discs help cardiac muscle cells to reproduce.

4. A neuromuscular junction is the location where a nerve tells a muscle what to do.

5. FOP is a disorder that causes muscles to turn into bone.

(Answers are on page 127.)

Integumentary System

Functions

The skin, or integumentary system, obviously keeps our insides in and our outsides out, but it's much more important than that! The skin, along with its appendages, hair and nails, makes up the largest organ in the human body. While it is thicker in some body regions than in others or has more hair or sweat glands in some parts than in others, the skin in general is pretty consistent as it covers the entire body. The skin is our first line of defense against pathogens and chemicals that can make us sick. The skin also functions in temperature regulation and maintaining homeostasis. The skin helps to manufacture vitamin D; stores fats, salts, glucose, and water; and has many nerve endings to help the body interpret its environment.

PAINLESS FACT

One average square centimeter of skin contains three million cells, 12 feet of nerves, three feet of blood vessels, 10 hairs, and 700 sweat glands!

Microscopic

The skin is made up of three layers: the innermost, or hypodermis (or subcutaneous); the middle, or dermis; and the outermost, or epidermis. The subcutaneous layer is composed of fat and loose connective tissue. Its function is to attach the skin to the surface muscles below. The dermis is the thickest layer of skin and is made up of collagen tissue, elastic fibers, dense connective tissue, blood vessels, muscle fibers, oil and fat glands, hair follicles, white blood cells, fat cells, and some mast cells. There are many nerve receptors in the dermis, each responding to specific stimuli, including response to pain, strong and light pressure, heat, cold, and general touch. Blood vessels in the dermis constrict (tighten) or dilate (expand) to adjust blood flow to the body surface and, as a result, regulate body temperature. The epidermis is the layer with keratin, used to waterproof and toughen the outermost skin. The epidermis also contains cells that produce melanin, which colors the skin and protects it from the sun's ultraviolet rays. There are no blood vessels in the epidermal layer. As epidermal cells get closer to the outside of the body, they begin to die off. As a matter of fact, when you look at yourself (or anyone else) the skin that you can see is made up entirely of dead cells!

HUMAN SKIN ANATOMY

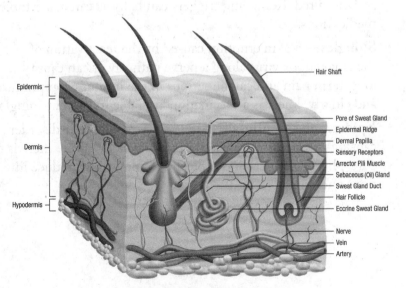

Figure 6–8. Cross Section of the Skin

Macroscopic

The skin is a body system that primarily functions at the macro-scopic level. Gross anatomy of the skin is pretty consistent across the entire body. Appendages of the skin, such as hair, sweat glands, pores, and nails, are also structures of the gross anatomy. Hair can function to help protect the body against the sun's rays and against some trauma; for example, eyelashes help to prevent small particles from getting into the eye. Sweat glands and pores remove some waste products from the body and help to cool the body. Nails are important in protecting the tips of our fingers and toes.

Disease

Integumentary diseases and disorders are often very obvious because they are on the outside of the body. Some common integumentary issues include cuts and abrasions, burns, and acne. As in any body system, problems can arise from injury, genetic and birth defects, bacterial and fungal infection, or cancer, or they may be idiopathic. Here are a few examples of skin disorders:

- **Eczema**—An inflammatory skin disease that is not conta-gious and can be acute or chronic. It can be caused by many different things, including allergies. Skin becomes dry, itchy, scaly, and red. Removing triggers can help, as can using topical medications.

- **Shingles**—A skin eruption caused by the reactivation of the chickenpox virus along a nerve pathway. It can cause long-term pain and can be serious in the immunocompromised and elderly. There is a vaccination available to prevent shingles.

- **Epidermolysis bullosa (EB)**—A very rare genetic disorder causing extremely fragile and blistering skin. While there is no cure, pharmaceutical and topical treatments may reduce infec-tions and pain.

BRAIN TICKLERS   Set # 3

Name the term that represents each of the following descriptions.

1. The skin layer with dead cells, it keeps us waterproof and helps block the sun's ultraviolet rays

2. A layer with mostly fat and connective tissue

3. A layer of skin that contains many blood vessels, nerves, sweat glands, and pores

A. epidermis

B. dermis

C. hypodermis

(Answers are on page 127.)

Cardiovascular System

Functions

The cardiovascular system is made up of the heart, the blood, and all of the blood vessels. The heart is a pump; the blood is a liquid that carries gases, nutrients, and wastes to and from all of the cells; and the blood vessels are the tubes that carry the blood throughout the body.

Microscopic

While most of the cardiovascular system operates on a macroscopic level, the blood is observed as a microscopic tissue. Blood is the main transporting fluid of the body. It is made of plasma, a completely liquid blood component that also transports nutrients to the cells and cellular waste to be disposed of, and cellular elements, the blood solids. Blood solids include red blood cells, five types of white blood cells, and platelets. Red blood cells deliver necessary oxygen to the cells and pick up waste (carbon dioxide) to be removed in the lungs.

PAINLESS FACT

Red blood cells only live for about four months. In that time, they travel throughout the entire body approximately 250,000 times.

The different types of white blood cells provide a strong immune defense against infections and disease. Platelets assist in clot formation.

Macroscopic

The heart is an organ made of cardiac muscle containing four chambers, four valves, and four major blood vessels entering and exiting it. It is surrounded and protected by the pericardium, its visceral membrane. The four chambers consist of two superior atria, collection and holding areas for blood, and two inferior ventricles, or chambers that pump out the blood. Blood is held in each chamber by one-way valves that let the waiting blood through and then close to allow the chamber to refill. Healthy circulation through the heart moves continually in one direction.

PAINLESS FACT

The reason that mammals have four-chambered hearts is that they have two separate circulatory systems: one to the lungs to eliminate carbon dioxide and pick up oxygen, and the other to the body to distribute oxygen to the cells and pick up waste. In a healthy person these two circulations do not interact.

PAINLESS STEPS

Understanding cardiac circulation is painless!

1. Blood (low in oxygen and high in waste) enters the right atrium from the vena cava.

2. Blood passes through the tricuspid valve into the right ventricle.

3. Blood travels through the pulmonary semilunar valve into the pulmonary arteries.

4. Blood moves into capillaries surrounding alveoli in the lungs. Carbon dioxide waste is diffused into the alveoli and oxygen is diffused into the blood.

5. Highly oxygenated blood travels through the pulmonary veins back to the left atrium.

6. Blood passes through the mitral valve into the left ventricle.

7. The left ventricle pumps blood through the aortic semilunar valve.

8. Highly oxygenated blood leaves the heart through the aorta to distribute oxygen throughout the body.

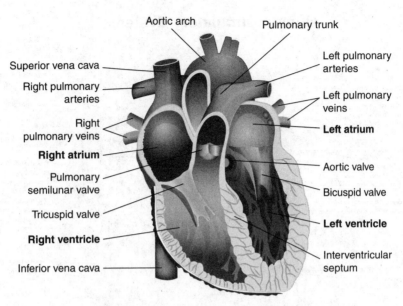

Figure 6–9. Heart Anatomy Cross Section

Circulation throughout the body occurs as the blood vessels transport oxygen in the blood from the heart to the cells. Arteries carry blood away from the heart and veins carry blood back to the heart. Blood is continually moving in one direction through the arteries because pressure exerted when the left ventricle contracts pushes blood through the arteries and into the capillaries. Capillaries are the thin-walled vessels in between that allow diffusion of oxygen into the cells and carbon dioxide waste from cells back into the blood. Capillaries then feed the blood into veins. Veins are not under pressure. Veins keep blood moving in one direction using a series of one-way valves that prevent backflow.

Human Circulatory System

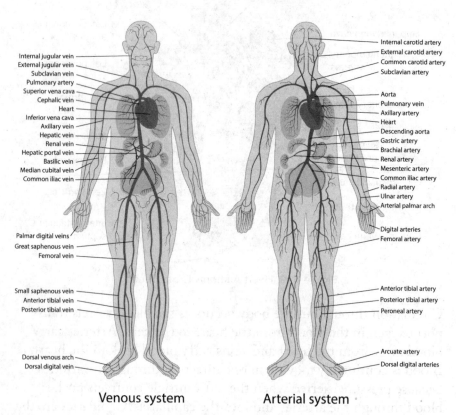

Internal jugular vein
External jugular vein
Subclavian vein
Pulmonary artery
Superior vena cava
Cephalic vein
Heart
Inferior vena cava
Axillary vein
Hepatic vein
Renal vein
Hepatic portal vein
Basilic vein
Median cubital vein
Common iliac vein

Palmar digital veins
Great saphenous vein
Femoral vein

Small saphenous vein
Anterior tibial vein
Posterior tibial vein

Dorsal venous arch
Dorsal digital vein

Internal carotid artery
External carotid artery
Common carotid artery
Subclavian artery

Aorta
Pulmonary vein
Axillary artery
Heart
Descending aorta
Gastric artery
Brachial artery
Renal artery
Mesenteric artery
Common iliac artery
Radial artery
Ulnar artery
Arterial palmar arch

Digital arteries
Femoral artery

Anterior tibial artery
Posterior tibial artery
Peroneal artery

Arcuate artery
Dorsal digital arteries

Venous system   Arterial system

Figure 6–10. Human Arterial and Venous Circulation

Disease

Some common cardiovascular disorders include hypertension, or high blood pressure, and anemia, a low red blood cell count. As in any body system, problems can arise from trauma, genetic and birth defects, infection, or cancer, or they may be idiopathic. Here are a few examples of cardiovascular disorders:

- **Hemophilia**—A sex-linked genetic disorder of the blood in which the blood clots abnormally or slowly. There are several forms of this disorder that can't be cured but can be treated with avoidance of trauma, medications, blood transfusions, and clotting factor.

- **Atherosclerosis**—Hardening of the arteries, a result of plaque accumulation in arterial walls. This can cause a blockage of blood flow to parts of the heart, causing a myocardial infarction. A healthy diet and exercise can help in prevention, and routine medical care can address the need for medications to minimize plaque by reducing blood pressure, cholesterol, and triglycerides in the blood.

- **Myocardial infarction (MI)**—Known as a heart attack, an MI is caused by a lack of blood flow to the heart muscle. Without blood supply, the muscle of the heart is damaged. Survival is very possible but depends on how quickly medical intervention occurs, the location of the tissue involved, how long the area was starved for blood, and the size of the area affected.

Respiratory System
Functions

The respiratory system consists of the organs related to breathing. The structures of this system include the lungs, bronchi, trachea, nose, and mouth. The rib cage and diaphragm assist in breathing. **Inspiration**, or breathing in, brings oxygen to the blood vessels in the lungs and **expiration**, or breathing out, gets rid of carbon dioxide. Oxygen, received through inspiration, enters the bloodstream and is sent to cells throughout the body.

Microscopic

The smallest units of the respiratory system are the alveoli. They are the endpoints of the tubes that make up the lungs and are surrounded by capillaries. As breathed air enters the alveoli, the oxygen, in high concentration, diffuses into the blood in surrounding capillaries. At the same time, carbon dioxide diffuses from the blood, where it's in high concentration, into the alveoli to be breathed out.

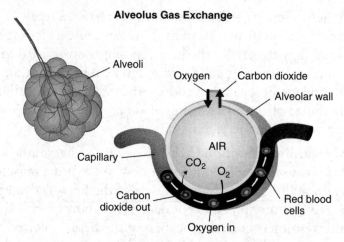

Alveolus Gas Exchange

Figure 6–11. Respiratory Gas Exchange

Macroscopic

Human respiration involves taking in air through the nose or mouth, moving it through the pharynx (back of the throat), past the larynx (voice box), down the trachea and the bronchi, and into the bronchioles, ending in the alveoli. At the alveoli, the oxygen in the air diffuses into the bloodstream to transport it to all the cells of the body. Equally important is the reverse process, which carries waste carbon dioxide that is diffused out of the blood and then follows the reverse path to be expelled through the nose or mouth.

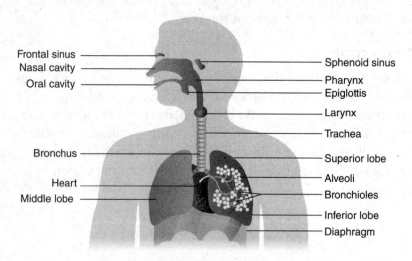

Figure 6–12. The Respiratory System

Disease

Some common respiratory disorders include the common cold, the flu, and rhinitis (inflammation of the nasal mucous membranes). As in any body system, problems can arise from trauma, genetic and birth defects, infection, or cancer, or they may be idiopathic. Here are a few examples of respiratory disorders:

- **Asthma**—An inflammatory disorder of the lungs that causes obstruction of the airways. Wheezing and difficulty exhaling are common in asthma patients. Medications that open and relax the airways usually help.

- **Pneumonia**—An infection of the lung, pneumonia can be caused by bacteria, viruses, or fungi. The infection causes the alveoli to fill up with a thick fluid that contains pus. Difficulty breathing and chest pain are symptoms. It is usually treated and resolved with medications and supplemental oxygen.

- **COPD**—A group of chronic lung diseases, including emphysema, that reduce airflow. It is often triggered by environmental factors like smoking, pollutants, and chemical exposures. In COPD, alveoli may become less elastic, which results in progressive difficulty in breathing.

BRAIN TICKLERS   Set # 4

Name the term that represents each of the following descriptions.

1. Pulmonary structures that allow exchange of gases

2. A waste product of cellular metabolism

3. The hearts holding chambers for blood

4. Chambers of the heart that pump blood

A. atria

B. ventricles

C. alveoli

D. carbon dioxide

(Answers are on page 127.)

Digestive System
Functions

The digestive system is the body system that takes in nutrients, breaks them down into molecules that the body can use, and then gets rid of any remaining waste. This system is made up of a single tube called the alimentary canal, which begins at the mouth and continues through the pharynx, esophagus, stomach, small intestine, large intestine, and anus. The digestive system also includes some accessory organs, such as the salivary glands, liver, gallbladder, and pancreas.

Microscopic

The goal of the digestive system is to break down nutrients so that they can be absorbed. Absorption of nutrients primarily happens in the small intestine wall, in fingerlike projections called villi. Villi contain embedded blood and lymphatic capillaries (called lacteals) that absorb digested food and pass nutrients on to the cells. Any remaining waste material is passed into the large intestine.

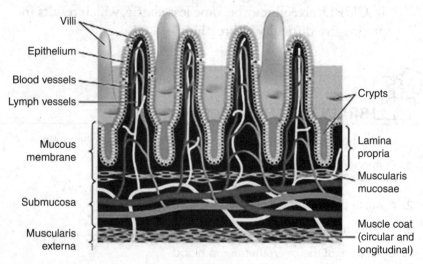

Figure 6–13. Cross Section of Small Intestine Wall

Macroscopic

As food moves through the organs of the digestive system, both mechanical and chemical forces help it to break down. Mechanical forces include chewing and grinding in the mouth, and muscular contractions throughout the rest of the digestive system. Each segment of this system plays a different role in the digestion process.

PAINLESS STEPS

Understanding the process of digestion can be painless!

1. The front teeth (incisors) slice off a chunk of food. Canine teeth will help if you need to tear something. The molars with their flat, bumpy surfaces will begin grinding the food. The three pairs of salivary glands will secrete saliva, which will lubricate the food and add enzymes to begin digesting any carbohydrates.

2. The food is now a chewed-up clump of material called a bolus, ready for you to swallow.

3. The bolus passes the pharynx and enters the esophagus where waves of muscular contractions, called peristalsis, begin. Peristalsis moves the food to the bottom of the esophagus.

4. The bolus, pressing on the hiatal sphincter, causes the sphincter to open, allowing the bolus into the stomach.

5. In the stomach, the layers of gastric muscles contract in different directions, mushing the food around. Enzymes and hydrochloric acid are secreted to begin digesting proteins.

6. After a few hours the food is liquified; it has a pH of 2 and is called chyme.

7. Chyme triggers the opening of the duodenal sphincter, moving it into the beginning of the small intestine.

8. In this section of the small intestine (the duodenum), the gallbladder and pancreas secrete chemicals to begin the digestion of fats and to bring the pH back up closer to neutral.

9. As peristalsis moves the nutrients along, the lacteals and blood vessels throughout the small intestine absorb most of the nutrients from the food.

10. At the end of the small intestine, what is left is mostly waste and water.

11. This material passes into the large intestine, where most of the water is reabsorbed, forming solid waste called feces.

12. Feces move into the final section of the large intestine, the rectum, to await release from the body at the anus.

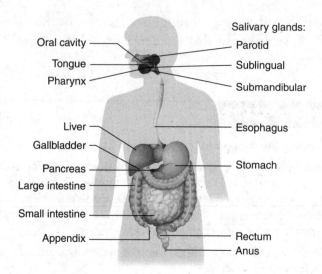

Figure 6–14. The Human Digestive System

Disease

Some common digestive disorders include appendicitis, gastro-enteritis (inflammation of the membranes lining the stomach and intestines), and heartburn. As in any body system, problems can arise from trauma, genetic and birth defects, infection, or cancer, or they may be idiopathic. Here are a few examples of digestive disorders:

- **Hepatitis**—Several different viral infections that cause liver inflammation. Vaccinations are available for some strains.
- **Inflammatory bowel disease (IBD)**—Crohn's disease and ulcerative colitis; both are autoimmune diseases that cause many uncomfortable symptoms, particularly chronic diarrhea.
- **Cholecystitis**—An inflammation of the gallbladder causing blockage of the bile duct. Often caused by gallstones.

Urinary System

Functions

The urinary system eliminates toxins that circulate in the blood. Elimination of toxins is accomplished by highly specialized microscopic filtration units in the kidneys. This system consists of the kidneys, ureters, urinary bladder, and urethra.

Microscopic

Each of our two kidneys contains over a million filtration structures called nephrons. Each nephron begins with a cuplike Bowman's capsule that holds a capillary cluster called a glomerulus. The capillary contains blood that needs to be rid of metabolic waste. The waste filters from the capillary into Bowman's capsule and then moves through the proximal tubule, the loop of Henle, and the distal tubule, and into a collecting tubule. Throughout this process, there is movement of molecules into and out of the tubules and water is added. By the time this filtrate reaches the collecting duct, toxins should have been removed and the pH adjusted, and the percentage of water should have been modified to retain necessary water while diluting toxins enough to continue safely through the system.

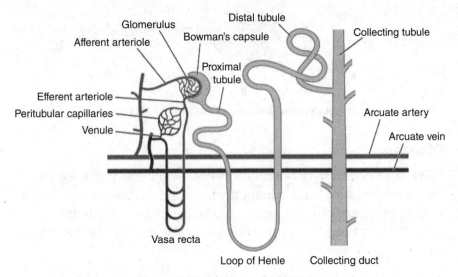

Figure 6–15. Anatomy of a Nephron

Macroscopic

When the nephron's filtrate leaves the collecting tubule, it collects in the base of the kidney and is called urine. The urine travels through the ureters and is stored in the urinary bladder. As the urinary bladder fills and expands, the pressure increases and ultimately the urine is released through the urethra. The urethra has two round sphincters (circular muscles) to help control the release of urine and prevent leakage.

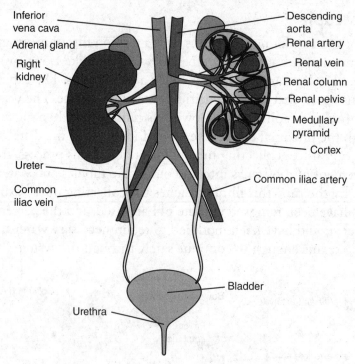

Figure 6–16. Anatomy of the Urinary System

Disease

Some common urinary system disorders include incontinence (involuntary urination) and urinary tract infections. As in any body system, problems can arise from trauma, genetic and birth defects, infection, or cancer, or they may be idiopathic. Here are a few examples of urinary disorders:

- **Kidney failure**—Nephrons no longer work effectively. Kidney failure causes reduced urine formation, which results in blood toxins accumulating in the body. This disorder can be acute or chronic. When severe, hemodialysis (mechanical blood filtration) or a kidney transplant are required.

- **Polycystic kidney disease (PKD)**—A genetic disorder that causes cysts to form throughout the kidneys, causing reduced kidney function. Often results in high blood pressure and kidney failure. Most patients eventually need dialysis or a kidney transplant.

- **Glomerulonephritis**—Caused by inflammation of the glomerulus, this disorder impacts the filtration process. It can be acute or chronic and often allows proteins and red blood cells to be retained and released in the urine. Some cases can be cured with medications, while other cases result in kidney failure.

BRAIN TICKLERS   Set # 5

Put the following processes in order as they occur in the human body.

A. Nutrients are absorbed into the bloodstream.

B. Peristalsis moves a bolus of food through the esophagus.

C. Food is chewed and carbohydrates begin digesting.

D. Waste is eliminated from the body.

E. Food is liquified into chyme and enzymes begin to digest proteins.

F. Water is reclaimed from remaining waste.

(Answers are on page 128.)

Nervous System

Functions

The nervous system is a complex and coordinated body system that dictates body functions and allows humans to think, to respond, and to recognize stimuli and adjust to them. Communication throughout the body and the ability to have the body perform quickly and collectively are functions of the nervous system.

PAINLESS FACT

Information can travel along neurons from the brain to the arms and legs at up to 268 miles per hour!

Microscopic

There are several types of nerve cells. A typical neuron has dendrites that receive a stimulus, either a physical stimulus or a chemical called a neurotransmitter, that travels in one direction through the dendrites toward the nerve cell body. From there the impulse is converted to an electrical signal and travels on through the axon, still in one direction, to the axon terminals, which send the neurotransmitter into the space, or synapse, at the end of the terminal. After crossing the synapse, the neurotransmitter is received by the next dendrite, moving the stimulus along to initiate the body's response. There are a variety of cell types that support the neurons without carrying impulses, including microglia, astrocytes, and oligodendrocytes. These glial cells protect, support, and insulate the neurons.

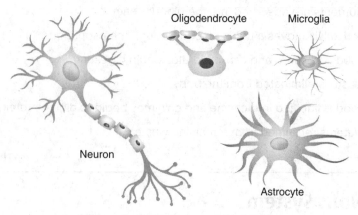

Figure 6–17. Neurons and Glial Cells

Macroscopic

The nervous system is divided into the central nervous system (CNS), containing the brain and spinal cord, and the peripheral nervous system (PNS), which is made of all the other nerves. The PNS is divided into sensory neurons, which take in information from our sensory organs, and motor neurons, which tell our muscles, glands, and organs what they need to do. The motor neuron system then divides further into the somatic (voluntary) and autonomic (regulating involuntary responses) nervous systems. Finally, the autonomic nervous system divides into the sympathetic division, which triggers the flight-or-fight response, and the parasympathetic division, which calms the body, bringing it back from the heightened intensity of the sympathetic response.

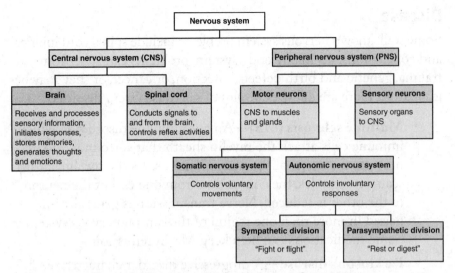

Figure 6–18. Levels of Nervous System Organization

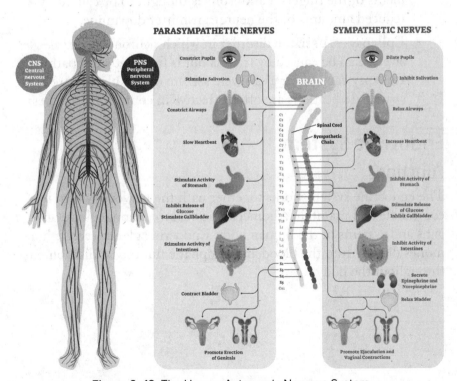

Figure 6–19. The Human Autonomic Nervous System

Disease

Some well-known nervous system problems include spinal cord injuries and concussions. As in any body system, problems can arise from trauma, genetic and birth defects, infection, or cancer, or they may be idiopathic. Here are a few examples of nervous system disorders:

- **Multiple sclerosis (MS)**—An autoimmune disorder in which immune cells attack the myelin sheath that surrounds and protects the axons of the body's neurons. As the myelin sheath scars or degrades, the connection from one end of the neuron to the other is reduced. Nerve transmission is generally impaired in more than one region of the central nervous system. Some medications can help keep MS in remission.

- **Parkinson's disease**—A progressive disorder characterized by tremors, shuffling walk, rigid muscles, and pill-rolling movements of the fingers. Parkinson's is thought to be caused by reduced amounts of the neurotransmitter dopamine.

- **Epilepsy**—A seizure disorder in which neurons fire excessively. Can be caused by trauma but is frequently idiopathic. Epilepsy can appear as a severe seizure or as a blank stare into space. Medications are often effective in controlling seizures.

Endocrine System

Functions

The endocrine system works together with the nervous system to coordinate and maintain a homeostatic balance among all the organs and systems of the human body. Endocrine glands secrete hormones directly into the blood and lymphatic fluid to be distributed throughout the body.

PAINLESS FACT

There is another type of gland called an **exocrine gland**. Exocrine glands secrete materials, both internally and externally, through tubes called ducts. Sweat glands release sweat through ducts in the skin and salivary glands secrete saliva through ducts inside the mouth.

Microscopic

All of the organs of the endocrine system produce and secrete hormones to affect other organs throughout the body. We produce many different hormones. Some of these hormones help us with digestion and metabolism (insulin, leptin, and ghrelin). Some help us to grow appropriately (human growth hormone). Some hormones direct sexual development and pregnancy (estrogen, follicle-stimulating hormone, oxytocin, and testosterone). Some hormones activate our fight-or-flight response (epinephrine) and some bring us back to normal after a scare (norepinephrine). One hormone helps us to adjust to the changing daylight hours throughout the year and can help us to sleep (melatonin).

PAINLESS FACT

The human body makes almost 30 different hormones. Other mammals also make essentially the same hormones and use them similarly to humans.

Macroscopic

The endocrine glands are the hypothalamus, pineal and pituitary glands in the brain, and the thyroid and parathyroid in the throat. The thymus gland is found under the sternum and above the heart. The adrenal glands sit atop each kidney. The pancreas lies behind the stomach. The gonads, or sex glands, include female ovaries in the pelvic cavity and male testes in the scrotum.

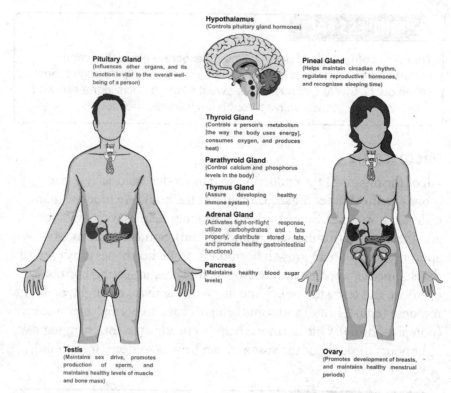

Hypothalamus
(Controls pituitary gland hormones)

Pituitary Gland
(Influences other organs, and its function is vital to the overall well-being of a person)

Pineal Gland
(Helps maintain circadian rhythm, regulates reproductive hormones, and recognizes sleeping time)

Thyroid Gland
(Controls a person's metabolism [the way the body uses energy], consumes oxygen, and produces heat)

Parathyroid Gland
(Control calcium and phosphorus levels in the body)

Thymus Gland
(Assure developing healthy immune system)

Adrenal Gland
(Activates fight-or-flight response, utilize carbohydrates and fats properly, distribute stored fats, and promote healthy gastrointestinal functions)

Pancreas
(Maintains healthy blood sugar levels)

Testis
(Maintains sex drive, promotes production of sperm, and maintains healthy levels of muscle and bone mass)

Ovary
(Promotes development of breasts, and maintains healthy menstrual periods)

Figure 6–20. Endocrine Glands and Their Functions

Disease

Most endocrine problems involve either hyperactivity of the glands, causing too much hormone secretion, or hypoactivity, resulting in the undersecretion of hormones. As in any body system, problems can arise from trauma, genetic and birth defects, infection, or cancer, or they may be idiopathic. Here are a few examples of endocrine disorders:

- **Diabetes mellitis**—Also known as type 1 diabetes, this disorder may be triggered by autoimmune, genetic, and viral factors that destroy the pancreatic cells that produce insulin. Cells need insulin in order to metabolize sugar. Sugar then accumulates in the blood rather than being used by the cells for energy. Without glucose for energy, the body begins to metabolize proteins and fat for energy. All these factors contribute to the body losing homeostatic balance. Blood glucose monitoring and replacement insulin are important in treating type 1 diabetes.

- **Hyperthyroidism**—The result of thyroid gland overactivity. Most often a result of an autoimmune disorder called Graves' disease. Symptoms may include swelling in the neck, bulging eyes, rapid pulse, increased blood pressure, weakness, and hair loss. Usually treated successfully with medication.

- **Cushing's syndrome**—Results from hypersecretion of adrenal hormones. Can cause weakness, high blood pressure, poor wound healing, and obesity. Surgical removal of part of the adrenal gland may help resolve this syndrome.

BRAIN TICKLERS   Set # 6

Name the term that represents each of the following descriptions.

1. Fight-or-flight response

2. Organs that secrete hormones directly into blood

3. A disorder caused by destruction of pancreatic cells

4. Cells that support neurons

A. glial cells

B. sympathetic nervous system

C. endocrine glands

D. diabetes mellitus

(Answers are on page 128.)

Lymphatic and Immune Systems
Functions

The lymphatic and immune systems work together to maintain homeostasis and fight illness and infection. Both systems share organs and produce cells that protect the body from pathogens and parasites.

Lymphatic system

The lymphatic system acts as a supplementary circulatory system in our bodies. It is made up of lymph nodes and vessels, but does not contain a pump like the circulatory system's heart. While our healthy circulatory system contains our blood at all times, we do have interstitial fluid moving throughout our bodies, outside of vessels. Interstitial fluid is similar to blood plasma, and when it moves out of intercellular spaces and into the system of lymphatic vessels, it is called lymph. A healthy lymphatic system returns all excess interstitial fluid back to the circulatory system. The spleen, thymus, tonsils, and lymph nodes act to filter out pathogens and also produce types of white blood cells that help fight infections. Lacteals, lymphatic capillaries in the villi of the intestines, help with absorption of fats and fat-soluble vitamins.

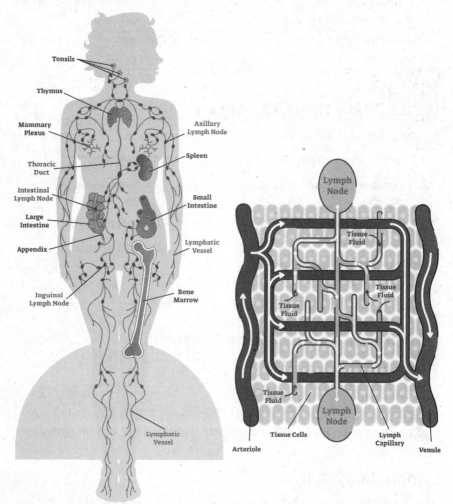

Figure 6–21. Structures of the Lymphatic System

Immune system

The job of the immune system is to protect the body from toxins, pathogens, allergens, and malignant (cancerous) cells. Natural immunity is the permanent and inherited immunity we have at birth.

The skin and mucous membranes provide the simplest and most basic protection as they block access to the inner body. Beyond that, many different types of antibodies, immunoglobulins, and white blood cells circulate through the body to protect it from invading pathogens. Acquired immunity is a form of immunity that is developed in response to environmental exposures. Passive acquired immunity is short term and is borrowed; an example is the temporary immunity passed to a newborn from its mother by way of the placenta and then from the mother's milk. Active acquired immunity is longer lasting immunity; it can come from having and recovering from some infectious diseases and from vaccinations.

PAINLESS FACT

Did you know that stress and lack of sleep can negatively impact your immune system? Similarly, positive emotions and a healthy lifestyle can boost immunity!

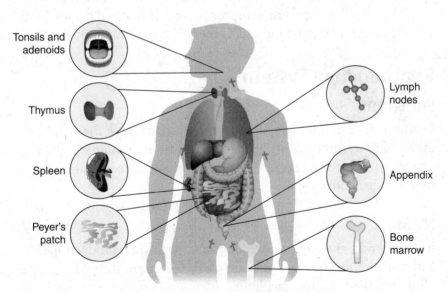

Figure 6–22. Structures of the Immune System

Disease

Most lymphatic disorders involve impaired circulation through the lymphatic system. Most immune disorders involve the immune system attacking its own body, and are called autoimmune disorders. As in any body system, problems can arise from trauma, genetic and birth defects, infection, or cancer, or they may be idiopathic. Here are a few examples of lymphatic and autoimmune disorders:

- **Lymphedema**—Swelling of body tissues because of accumulated lymphatic fluid. It is caused by damage to the lymphatic system that keeps lymph from draining correctly. Treatment is difficult but often includes specialized physical therapy.

- **Scleroderma**—An autoimmune disease that causes skin and blood vessels to become tough and thickened. Usually, people with scleroderma also have cold sensitivity in their fingers and toes. Treatment usually includes immune system suppressants and symptom relief.

- **Systemic lupus erythematosus**—The autoimmune disease lupus is a chronic and inflammatory immune system attack on its own body tissues. Symptoms include unusual rashes, joint pain, and extreme fatigue. Generally treated with medications.

Reproductive System

Functions

Both the male and female reproductive systems serve the same two major purposes: they contribute to the creation of unique offspring, and they produce and secrete hormones involved in sexual reproduction and development.

Male

Male reproductive organs include the testes, which produce sperm and testosterone. A series of ducts carry the sperm through the epididymis, vas deferens, ejaculatory ducts, and urethra. The urethra is also a part of the urinary system and is the structure that moves urine out of the body. Accessory organs to the male reproductive system are the seminal vesicles, bulbourethral glands, and prostate gland. These organs produce a fluid to nourish and help transport the sperm. The penis serves as a mechanism to transfer sperm cells to the female reproductive system.

Female

The female reproductive system includes the ovaries, which produce hormones including estrogen and are the structures where ova, or egg cells, develop. It also consists of the fallopian tubes, where fertilization of the egg generally occurs; the uterus, where the fertilized egg (zygote) implants and develops; and the vagina, where sperm is introduced into the female reproductive system and where a baby emerges during childbirth. Menstruation is a monthly physical and hormonal cycle that provides a fresh and healthy environment for the development of a fetus. Breasts are accessory organs for the production and delivery of milk to a baby.

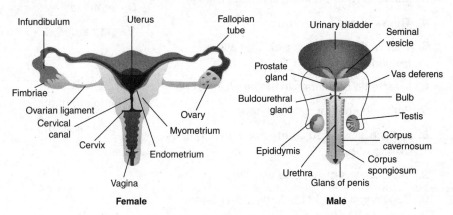

Figure 6–23. The Male and Female Reproductive Systems

Disease

- **Benign prostatic hypertrophy**—In males only, this is a noncancerous enlargement of the prostate gland. Frequently diagnosed in men over 60, it causes difficulty in emptying the urinary bladder.

- **Sexually transmitted diseases (STDs)**—A group of infectious diseases transmitted during sexual activity. STDs include chlamydia, human papillomavirus (HPV), gonorrhea, genital herpes, syphilis, and HIV/AIDS. Each of these can affect both males and females, and treatments vary.

- **Endometriosis**—In females only, this is a disorder in which the uterine lining, the endometrium, is found outside the uterus in the abdominopelvic cavity. This can cause inflammation of nearby tissues, pain, internal bleeding, and formation of scar tissue, and can result in fertility problems.

SUPER BRAIN TICKLERS

Match these definitions to the correct terms.

1. Waves of contractions that move material through the digestive system

2. The main contracting muscle in a pair of skeletal muscles

3. Location where an egg is fertilized

4. The skull and vertebrae

5. Thin-walled blood vessels that allow exchange of O_2 and CO_2

6. Lymphatic capillaries in the small intestine

7. Pumping chambers of the heart

8. A disorder caused by the body attacking its own tissues

9. Vaccinations provide us with this

10. Capillary cluster in the kidney that brings blood to be filtered

A. lacteals

B. capillaries

C. active acquired immunity

D. ventricles

E. glomeruli

F. peristalsis

G. fallopian tube

H. autoimmune

I. axial skeleton

J. agonist

(Answers are on page 128.)

Vocabulary

Anterior: Opposite of posterior; anterior is the front side of the body.

Articulation: The place where two bones come together; a joint.

Distal: When referring to the arms and legs, distal describes the areas farthest from the torso.

Dorsal cavity: A body space in humans that contains the cranial cavity (brain) and the vertebral cavity (spinal cord).

Expiration: Breathing out.

Frontal: A plane of dissection in the human body that separates the front (anterior) from the back (posterior) of the body.

Gross anatomy: Tissues, organs, and systems are generally easier to see without a microscope and are called macroscopic, or **gross anatomy**. The word *gross* here means large enough to be seen with the naked eye.

Hematopoiesis: The process of stem cells differentiating into red and white blood cells and platelets in the red bone marrow.

Idiopathic: Disease that occurs with no known cause.

Inferior: Directional term for parts of the body toward the feet.

Insertion: Mobile point(s) of attachment of a skeletal muscle to bones.

Inspiration: Breathing in.

Intercalated disc: Structures in cardiac cells allowing them to have a unique ability to communicate with each other and to create a normal rhythmic heartbeat.

Lateral: A directional term referring to body parts located away from the midsagittal line, which divides the right and left sides.

Ligament: The tough connective tissue that holds one bone to its adjoining bone.

Medial: A directional term referring to body parts located toward, or close to, the midsagittal line, which divides the right and left sides.

Neuromuscular junction: Location where a nerve intersects with a muscle, capable of triggering a muscular contraction.

Oblique: Angled plane of dissection.

Origin: Nonmoving point(s) of attachment of a skeletal muscle to bones.

Oxygen debt: Following intense exercise, the body may experience a deficit of oxygen needed for respiration, called an oxygen debt.

Anaerobic respiration covers this deficiency until the body recuperates and resumes normal aerobic respiration.

Parietal membranes: Serous membranes that line the insides of the body cavities and are continuous with the visceral membranes.

Posterior: Opposite of anterior; posterior is the back side of the body.

Proximal: When referring to the arms and legs, proximal describes the areas closest to the torso.

Sagittal: A line of dissection that divides the right and left sides of the body.

Serous membranes: Smooth tissue linings throughout the body that secrete lubricating fluid. Their job is to insulate, protect, maintain position, and lubricate all of our organs.

Superior: Directional term for parts of the body toward the top of the head.

Transverse: A horizontal line of dissection that divides the top from the bottom of the body.

Ventral cavity: A space found in the torso, or trunk. The ventral cavity is divided in half by the muscular diaphragm. Above the diaphragm is the chest cavity and below it is the abdominal cavity.

Visceral membranes: Serous membranes that surround the organs in the body and are continuous with the parietal membranes.

Brain Ticklers—The Answers

Set # 1, page 91

1. C
2. A
3. B

Set # 2, page 98

1. False
2. True
3. False
4. True
5. True

Set # 3, page 101

1. A
2. C
3. B

Set # 4, page 107

1. C
2. D
3. A
4. B

Set # 5, page 113

1. C
2. B
3. E
4. A
5. F
6. D

Set # 6, page 119

1. B
2. C
3. D
4. A

Super Brain Ticklers

1. F
2. J
3. G
4. I
5. B
6. A
7. D
8. H
9. C
10. E

Chapter 7

Cellular Reproduction

Asexual and Sexual Reproduction

Do you remember the characteristics of living things from Chapter 1? One of the characteristics of life is reproduction. What did that make you think of? Maybe a neighbor's newborn baby or a new puppy or kitten? A newly hatched pet snake, perhaps? Does it make you wonder how your skin heals when you get a cut? Or why your grandmother's skin looks so different from yours? Does it make you wonder what controls the aging process? All of these considerations contain elements of the process of reproduction.

The first thing to consider when you're investigating the topic of reproduction is to establish whether you are talking about asexual or sexual reproduction. There are many different reasons for cells to reproduce. When you fall and skin your knee, your skin cells undergo a form of **asexual reproduction**, called **regeneration**, to heal your injury. When that newborn baby, pet snake, kitten, or puppy grows in size, its body cells have been reproducing asexually so that it can get bigger. When tulips begin to emerge from their bulbs in the springtime, you can see them growing in height and producing flowers as a result of the rapid asexual reproduction going on in their cells. Asexual reproduction is important because it generates brand new cells that are identical to the parent cell as well as to their sister cells. These identical cells are important in situations like growing or healing a cut on your skin. You want replacement cells to appear and behave just like the cells that were there before. When asexual reproduction works correctly, there is no variation in the features of the new cells. **Sexual reproduction**, on the other hand, is the process of **fertilization**, where two unique parents each contribute a single **gamete** (egg or sperm cell) to produce a **zygote**. Formation of those

two gametes is an important element of the process of sexual reproduction. Sexual reproduction provides variation within members of
the same species. Sexual reproduction allows individuals and populations to adapt to changes in their environment and is an important
element in the process of **natural selection**.

PAINLESS TIP

You'll see the prefix "*a*-" a lot in scientific reading—it means "without."
Asexual reproduction means reproduction without sexual interaction.

The Cell Cycle

Every living cell goes through different developmental stages. These
stages may last for very different amounts of time and may have
minor differences, but they are similar in all cells and are collectively
called the **cell cycle**. The cell cycle has two main phases: **interphase**
and **mitosis** (or **meiosis**).

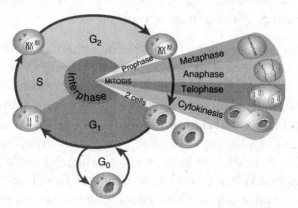

Figure 7–1. The Cell Cycle

The cell cycle begins with the formation of a new cell. That cell
enters the G1 phase of interphase. During G1 the cell will develop
and grow up; it will produce the organelles it needs to survive and
become a mature cell. It does the day-to-day work that its job requires. Once the cell is mature it has two options: it can continue on
as it has been by entering the G0 phase, or it can move toward the S
phase. The phase called G0 is a cycle of simply continuing on with

its day-to-day activities indefinitely—metabolizing nutrients, maintaining homeostasis, adding mass and organelles, staying organized, and continuing to mature. This phase is appropriate as long as the cell works efficiently and all the normal cell regulators, like cell-to-cell contact inhibition (when cells are prevented from reproducing as long as their borders are in physical contact with other cells), are functioning. Many of our nervous system cells stay in G0 for our whole lives, and our liver cells can remain in this phase for over six years. Our skin cells, on the other hand, are programmed to only live for about a month, so they'll stay in G1 or G0 for around 28 days and then move on to the **restriction point** in G1. Movement to the restriction point commits the cell to work toward the ultimate goal of cell division.

Immediately after reaching the restriction point the cell enters S phase (synthesis phase), where each single chromosome is duplicated. Each copy holds on to the original chromosome strand at a structure called a **centromere**, and together they are shaped like an "X" and called **sister chromatids**. They stay inside the nucleus for the time being. Once the process of forming sister chromatids is complete, the cell moves into the G2 phase. Here, the cell forms plenty of extra cytoplasm, membranes, and organelles in preparation for forming new cells during cell division. At the end of G2, the cell is ready to enter either **mitosis** or **meiosis**.

 BRAIN TICKLERS   Set # 1

Name the term that represents each of the following descriptions.

1. The structure that holds sister chromatids together   **A.** centromere

2. The process of replacing or restoring damaged cells   **B.** interphase

3. Egg and sperm cells   **C.** gamete

4. The first phases of the cell cycle   **D.** regeneration

(Answers are on page 141.)

Mitosis

If the cell is destined to undergo asexual reproduction, it enters mitosis at the end of G2. It begins with a stage called **prophase**. Essentially, the cell is being prepared for cell division. During prophase the sister chromatids are no longer a tangled cluster; they become clear and distinct inside the nucleus. The nuclear membrane begins to dissolve. The centrioles begin to move to opposite ends (poles) of the cell. As the centrioles move apart, they pull along strands called spindle fibers, which are important elements of the next stage of mitosis, **metaphase**. During metaphase, the nuclear membrane has disappeared, and the centromeres attach to spindle fibers and use them to line up in the equator of the cell. The next stage of mitosis is **anaphase**. Now, the centromeres split and the sister chromatids separate with the spindle fibers, pulling them to opposite poles of the cell. This stage is important because it ultimately places homologous alleles in two different new cells. Finally, the cell moves into a stage called **telophase** when steps are taken to divide the parent cell into two new, identical cells. During telophase a new nucleus forms around the cluster of chromosomes at the cell's poles and the spindle fibers dissolve. In animal cells, the large cell now begins to pinch together in the middle, separating the two new nuclei. Plant cells complete telophase differently; they form a cell plate along the equator, between the two nuclei. Finally, during **cytokinesis**, the parent cell separates into two new "baby" cells, all ready to enter the cell cycle and begin their own journey through interphase.

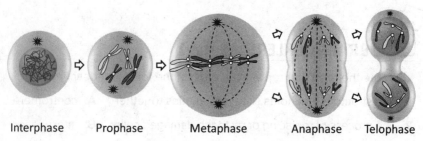

Interphase   Prophase   Metaphase   Anaphase   Telophase

Figure 7–2. The Stages of Mitosis

BRAIN TICKLERS   Set # 2

Name the term that represents each of the following descriptions.

1. Stage of mitosis when two new cells are ready to split apart

A. cytokinesis

B. metaphase

2. Stage of mitosis when sister chromatids become clear and distinct and the nuclear membrane begins to dissolve

3. Process during which two new "baby" cells are formed

4. Stage of mitosis when chromosomes are pulled to opposite poles

5. Stage of mitosis when sister chromatids line up along the cell's equator

C. prophase

D. telophase

E. anaphase

(Answers are on page 141.)

So, the bottom line regarding mitosis is that mitosis is the mechanism that all somatic (body) cells use to grow, develop, maintain homeostasis and organization, and conduct metabolism. Mitosis permits us to heal after we get a cut or a sunburn, or when we undergo surgery. Mitosis does its best to provide identical replacement cells for our damaged or aging body cells. The asexual process of mitosis is continuously occurring in our bodies and in all living cells and organisms.

PAINLESS TIP

If you're having trouble remembering the names or the sequence of the steps in mitosis, maybe it would help to remember the first letters and use an acronym. Prophase begins with "P" and is the step that "prepares" the cell to move forward with mitosis. Metaphase begins with "M" and is the step where sister chromatids line up in the "middle" of the cell. Anaphase begins with "A" and is the step when chromosomes move "away" or "apart." Telophase with a "T" is the time when the cell is splitting to form "two" new cells. How about the acronym PMAT? Many students take the PSAT before they take their SAT. Hopefully, this acronym will help you remember the meanings and sequence of prophase, metaphase, anaphase, and telophase.

Meiosis

If the goal of cellular reproduction is the production of offspring, humans and most other complex animals and plants use the process of sexual reproduction, which is initiated by **meiosis**. Meiosis creates gametes from diploid cells. In other words, meiosis begins with **diploid** (or 2n) cells that have a pair of homologous chromosomes (or alleles). Over the course of meiosis I and meiosis II, those alleles are evenly distributed into egg and sperm cells that are **haploid** (1n), or each containing one of the alleles for a specific chromosome pair. When fertilization occurs, an egg and sperm are united, yielding a zygote with a unique combination of alleles. Meiosis creates an opportunity for variation within a species and for favorable traits in an environment to be passed on to enable natural selection to occur.

Meiosis

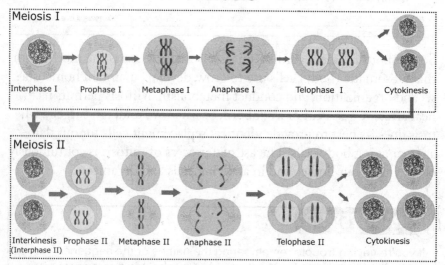

Figure 7–3. The Stages of Meiosis

You may be thinking that meiosis just looks like the cell going through mitosis twice. That's true, but there are some very important differences. The final result of mitosis is two diploid and identical daughter cells, while meiosis produces four haploid gametes, either egg or sperm cells. The products of mitosis, being identical to their parent and sister cells, can replace growing, aging, damaged, or dying body cells. They should be exact copies, so their appearance and function should be identical to their parent cells. The products of meiosis will be available to contribute half the chromosomes to a new and different offspring.

Other than going through two rounds of divisions, and even though the stages of PMAT have similar functions, there are significant differences between mitosis and meiosis. First, keep in mind that mitosis goes through PMAT while meiosis undergoes PI (pronounced prophase one), MI, AI, TI, then cytokinesis I, then PII, MII, AII, TII, and cytokinesis II. In meiosis during prophase I, every pair of homologous chromosomes lines up next to each other in a process called **synapsis** to form groups of four alleles (called tetrads). During this stage, genetic material is likely to be recombined between the alleles in a process called **crossing over**. In metaphase I the tetrads line up at the cell's equator, which allows DNA from each parent to line up on both sides of the midline. During anaphase I, sister chromatids remain attached to each other, but the homologous chromosomes (tetrads) are separated and moved toward opposite poles. In some species during telophase I, temporary nuclear membranes form around sister chromatids. When the cell enters cytokinesis I, the cell splits, becoming two cells each with a full set of chromosomes.

BRAIN TICKLERS   Set # 3

Name the term that represents each of the following descriptions.

1. Cells that are 2n (have a pair of homologous alleles)

2. Homologous sister chromatids line up and exchange DNA with each other

3. Cells that are 1n, with one of the alleles from a chromosome pair

4. Pairing of homologous chromosomes during prophase I, a stage when genetic recombination occurs

A. synapsis

B. haploid

C. crossing over

D. diploid

(Answers are on page 141.)

During the stages of meiosis II, the stages are a lot like the stages of mitosis. The difference is that at the end of mitosis there are two complete and identical diploid daughter cells, while at the end of meiosis II there will be four unique and haploid potential gametes (egg or sperm cells). The process of forming an egg cell continues on as three of the four newly formed egg cells become **polar bodies**, donating their cytoplasm to the one egg cell out of the four that will ultimately become a viable egg. So, each completion of meiosis produces four viable sperm cells or one viable egg cell.

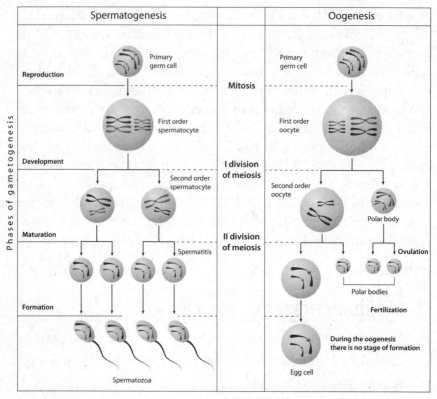

Figure 7–4. Gametogenesis: The Formation of Egg and Sperm Cells

Telomeres

You may be wondering, if cells can just keep copying themselves when they get old, injured or sick, why don't they just make a fresh copy and go on? You may be wondering, why would cells even die if mitosis can keep them alive? This would be called **cellular immortality** and is caused by uncontrolled mitosis. Immortal cells no longer follow the cell's rules for reproduction and can live for much longer than they should. Cellular immortality and uncontrolled mitosis can cause cancer.

Cells have a genetically predetermined life span, and they are programmed to die when they get to a specific age. Programmed cell death is called **apoptosis**. The mechanism that allows apoptosis is an end cap on both ends of every chromosome. The end cap is made of a series of repeating DNA nucleotides called **telomeres**. The telomeres shorten with each round of cellular reproduction. As a result, there are a limited number of times that most cells can divide. As the telomeres get smaller, there are more opportunities for mistakes to be made in copying and less protection of accurate cellular reproduction. Ultimately, when telomeres are too small or are missing, a lot can go wrong; mistakes in copying DNA, cellular aging, and death can occur. When they are working correctly, telomeres control how many times a cell reproduces. An enzyme called telomerase adds the correct DNA sequence to the ends of chromosomes to continue accurate programmed cellular reproduction. The aging process reduces telomerase activity and, by adulthood, most cells do not exhibit much telomere replacement. There are several types of adult cells that do continue to have telomerase activity throughout adulthood, including hair follicles, gamete producing cells, stem cells, and most cancer cells.

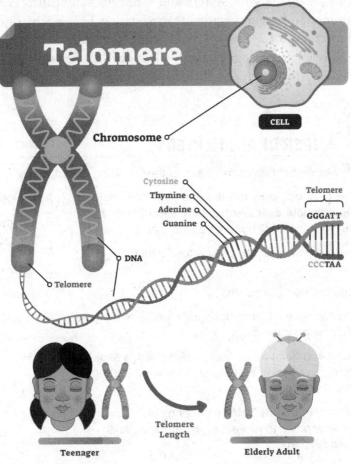

Figure 7–5. Aging Process as Telomeres Shorten

PAINLESS FACT

Have you ever noticed the difference between the skin of a baby and the skin of an elderly person? The baby usually has smooth, consistent skin while an older person may have spots or bumps causing a more irregular surface. This is because the baby has longer telomeres protecting the skin cells from variations and mistakes when they reproduce. As the baby ages through adulthood, their telomeres are no longer strong enough to prevent variations or mistakes from occurring, so daughter cells will not necessarily be identical to each other or to the parent cell. This causes variations in the skin like skin tags and age spots.

Telomere length and telomerase activity are genetically programmed for each type of cell in every species. We talked earlier about cellular life spans for the nervous system and skin cells. All healthy cells are programmed to have a specific life expectancy. This is why mice have a life expectancy of 2–3 years, many small dogs have a life expectancy of 10–15 years, large dogs generally live for 8–12 years, and a Galapagos tortoise may live for over 200 years!

SUPER BRAIN TICKLERS

Name the term that represents each of the following descriptions.

1. One of three newly formed egg cells from meiosis II that donate their cytoplasm to a fourth egg cell that will ultimately become a viable egg

2. The process of forming gametes for sexual reproduction

3. Programmed cell death

4. End cap on each chromosome; it protects the DNA while being copied

5. Asexual cellular reproduction that creates two identical offspring that are also identical to their parent cell

6. Enzyme that adds DNA to the ends of chromosomes to protect them during cellular reproduction

A. meiosis

B. telomerase

C. polar body

D. telomere

E. apoptosis

F. mitosis

(Answers are on page 141.)

Vocabulary

Anaphase: The third stage of mitosis, when the centromere splits and the sister chromatids separate, with spindle fibers pulling them to opposite poles of the cell.

Apoptosis: Genetically programmed cell death.

Asexual reproduction: Generates brand new cells that are identical to the parent cell as well as to its sister cells.

Cell cycle: The series of stages in a cell's life, basically comprised of interphase and mitosis.

Centromere: The structure that connects two sister chromatids to each other and to a spindle fiber during metaphase in mitosis.

Crossing over: During meiosis prophase I, homologous sister chromatids line up and exchange DNA with each other. This process allows variations to occur within both families and species.

Cytokinesis: At the end of mitosis, a parent cell has separated into two new "baby" cells each ready to enter the cell cycle and begin their own journey through interphase.

Diploid: Cells with two homologous chromosomes, represented by 2n.

Fertilization: The uniting of egg and sperm cells, which produces a single unique organism with genetic information from two parents.

Gametes: Egg and sperm cells, each containing one of the alleles for each pair of chromosomes that the offspring will have.

Haploid: Cells with one of the homologous chromosomes from a chromosome pair, represented by 1n.

Immortality: Cellular immortality is caused by uncontrolled mitosis. Immortal cells no longer follow the cell's rules for reproduction and can live for much longer than they should. Cellular immortality and uncontrolled mitosis can cause cancer.

Interphase: The growth, development, and DNA synthesis phase of the cell cycle.

Meiosis: The process of forming gametes for sexual reproduction.

Metaphase: The second phase of mitosis, where the nuclear membrane is gone and centromeres attach to spindle fibers and use them to line up in the center of the cell.

Mitosis: Asexual cellular reproduction that creates two identical offspring that are also identical to their parent cell.

Natural selection: The survival and reproductive advantages of individuals in a population as a result of differences in phenotype.

Polar bodies: Three of the four newly formed egg cells following meiosis II that donate their cytoplasm to the single egg cell that will ultimately become a viable egg.

Prophase: The first phase of mitosis, in which the nuclear membrane begins to disintegrate, the sister chromatids begin to be clear and distinct, and the centrioles begin moving to opposite poles of the cell.

Regeneration: The process of restoring or replacing damaged or missing cells or body parts in a living thing.

Restriction point: A stage in G1 of mitosis; movement to the restriction point commits the cell to work towards an ultimate goal of cell division.

Sexual reproduction: The process of fertilization, where two unique parents each contribute a single gamete (egg or sperm cell) to produce a zygote.

Sister chromatids: Formed during S phase, a copy of each chromosome is attached to its copy at the centromere.

Synapsis: Pairing of homologous chromosomes during prophase I, a stage when genetic recombination occurs.

Telomeres: The caps at both ends of a chromosome. Made of a series of repeating DNA nucleotides, these structures shorten with each round of cellular reproduction. Ultimately, when telomeres are too small or are missing, cellular aging and death occur.

Telophase: The fourth and final stage of mitosis, when steps are taken to divide the parent cell into two new, identical cells.

Zygote: The earliest stage of a fertilized egg cell.

Brain Ticklers—The Answers

Set # 1, page 131

1. A
2. D
3. C
4. B

Set # 2, pages 132–133

1. D
2. C
3. A
4. E
5. B

Set # 3, page 135

1. D
2. C
3. B
4. A

Super Brain Ticklers

1. C
2. A
3. E
4. D
5. F
6. B

Chapter 8

Basic Genetics

What Is Genetics?

Genetics is a huge and important field of study in the science of biology these days. It is concerned with **heredity**, the inheritance of information and traits from one generation to the next. The modern science of heredity and genetics goes back to Gregor Mendel's discoveries in the 1860s. However, humans were involved in modifying other species' heredity long before Mendel's discoveries. Humans were using a process called **selective breeding** during prehistoric times between 6,000 and 10,000 years ago. Humans domesticated crops (like maize, wheat, and millet) and animals (initially dogs, goats, and sheep) to better serve them. Selective breeding involved limiting and introducing reproductive opportunities for these plant and animal species to other plants or mating partners that had traits that were useful to humans.

To develop a domesticated crop, early farmers discovered and isolated certain maize (corn) plants that met their criteria for ease in planting and harvesting and that tasted best. These early humans understood, through their experience, that those plants would tend to produce offspring with similar traits and would be similarly tasty and easy to farm. Over time, selective breeding made these crops increasingly domesticated. It's likely that the earliest domesticated dogs were willing to live near people. They were probably able to provide some level of protection to the people they lived with. The humans, in return, gave them food and possibly shelter. As time and generations went by, dogs and humans became more interdependent, and dogs became more and more domesticated. During those thousands of years, without understanding the science behind it, early humans were recognizing and modifying the inheritance patterns of the plants and animals that they needed in order to be more successful in their environments.

So, what exactly is genetics, how is it related to selective breeding, and how does it work? Genetics is the science of understanding how and why inheritance patterns happen. We now recognize that genetics is a framework for understanding the passage of a trait from a parent to its offspring, and then to the offspring's progeny, and on and on for generation after generation. As scientific technology and advances have developed, humans have been better able to understand inheritance on an increasingly molecular and predictable level. We now understand that DNA in our cells contains information that codes for the **traits** that our bodies have. We receive half of our DNA instructions from the egg and half from the sperm that created each of us. This inherited DNA makes us similar to each of our parents, but also different from each of them.

What Are Traits and How Do They Work?

A trait is a very specific characteristic of an organism. Traits are all of the characteristics that make us who we are. They have variations, or differences, because of the blending of genes during fertilization and because of environmental influences throughout our lives. One of the things we look at in genetics is how traits are exhibited by different members of a family.

PAINLESS FACT

Can you curl your tongue? Does your hairline create a point, or widow's peak, in the middle of your forehead? Are your earlobes unattached and free at the bottoms of your ears? If you interlock your fingers, does your left thumb naturally land on top? Each of these is a trait that you have inherited from one or both of your parents!

Often, when scientists want to explore the inheritance patterns of traits, they use a model called a **pedigree**. We can follow a trait through several generations of parents and offspring to begin to understand how the trait is inherited. In a pedigree, males are represented by squares and females by circles. The shapes are colored in based on whether the trait is seen in that member of the family. Horizontal lines represent members of the same generation and vertical lines show us the offspring of the couple shown directly above them. Their mates have a horizontal line attaching them and a vertical line between them showing us their offspring. In Figure 8–1 below, the top two shapes are the original parents—we call them the parental, or P, generation. The next line displays their offspring and the partners of the offspring; these are called the first filial, or F1, generation. The term *filial* refers to sons and daughters. The third line down contains the offspring of the F1 generation; they are the grandchildren of the P generation. These offspring are called the second filial, or F2, generation.

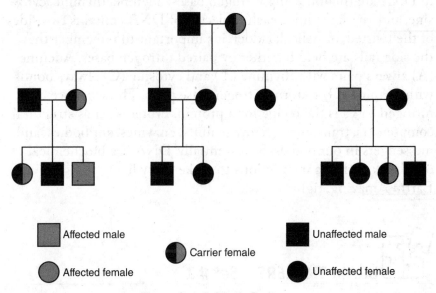

Figure 8–1. A Family Pedigree

PAINLESS TIP

Do pedigrees seem interesting to you? It can be a fun and interesting experience to select one of the traits mentioned in the last Painless Tip (tongue curling, widow's peak, earlobe attachment, and thumb dominance) and see the inheritance pattern in your own family. Obviously, families are different and sometimes we don't have access to family members to get the necessary information about their traits. You can practice making pedigrees using made-up information. And then, what if you have a trait that neither of your parents have? It happens a lot! How can that be, you ask? We'll be covering that in Chapter 9!

What Is DNA and How Does It Work?

Remember how we discussed DNA in Chapter 3? It never hurts to review, though, so let's explore DNA and its relationship to genetics. DNA, also known as deoxyribonucleic acid, is the spiral ladder–shaped molecule in our bodies that carries the instructions to make each of us. It's like the cookbook for you and me and every other living thing on Earth. The DNA "cookbook" is kept in a very specific order by the side rails of the DNA ladder. The side rails are identical in all living things. Throughout all strands of DNA, side rails are made up of the same sugar, deoxyribose, alternating with an identical phosphate group. The only variables in DNA are the following nitrogen bases: adenine, thymine, cytosine, and guanine. Since each molecule of DNA contains two sides of the twisted, or helical, ladder, it's important to remember that the side rails are held together by paired nitrogen bases. Adenine (A) always pairs with thymine (T) and cytosine (C) always bonds with guanine (G) to form nitrogen base pairs. The sequence of nitrogen bases is the recipe for a protein. Proteins act as structural components, transporters, storage units, enzymes, antibodies, and messengers in our bodies. So, essentially, DNA is a biochemical cookbook with the instructions to make every living thing on Earth! Amazing, right?

 BRAIN TICKLERS   Set # 1

Select the best term to complete each sentence.

1. Prehistoric humans worked to develop animals and crops that were most useful to them. They did this by using (<u>natural</u> or <u>selective</u>) breeding.

2. A very specific characteristic of an organism is a (<u>gene</u> or <u>trait</u>).

3. The instructions to make all living things are held in molecules called (<u>proteins</u> or <u>DNA</u>).

(Answers are on page 152.)

What Are Chromosomes and How Do They Work?

In bacteria, DNA is present in a single, usually circular, strand floating in the cytoplasm. In eukaryotes, DNA is usually found in linear strands folded, coiled, and contained in each cell's nucleus. Both types of strands are long pieces of DNA that are called **chromosomes**. Each chromosome contains the instructions for many of the traits and proteins that make up the organisms that contain them. Each species has a unique number of chromosomes. Ordinarily, every organism of a sexually reproductive species has two copies of each chromosome creating a chromosome pair. The term **diploid** means that the chromosomes are paired. In preparation for sexual reproduction, or meiosis, the paired chromosomes separate to provide the sex cells (egg and sperm) with a single chromosome from each pair. With that single chromosome from each chromosome pair, these gametes are called **haploid**. Fertilization with two haploid gametes will then result in unique diploid offspring. It's interesting that genetic modification has created an octoploid strawberry, with eight copies of each chromosome; this provides the strawberries with the structural materials to be much larger than unmodified strawberries.

Numbers of Pairs of Chromosomes in Different Species of Plants and Animals					
Common name	Species	Number of chromosome pairs	Common name	Species	Number of chromosome pairs
Mosquito	*Culex pipiens*	3	Wheat	*Triticum aestivum*	21
Housefly	*Musca domestica*	6	Human	*Homo sapiens*	23
Garden onion	*Allium cepa*	8	Potato	*Solanum tuberosum*	24
Toad	*Bufo americanus*	11	Cattle	*Bos taurus*	30
Rice	*Oryza sativa*	12	Donkey	*Equus asinus*	31
Frog	*Rana pipiens*	13	Horse	*Equus caballus*	32
Alligator	*Alligator mississipiensis*	16	Dog	*Canis familiaris*	39
Cat	*Felis domesticus*	19	Chicken	*Gallus domesticus*	39
House mouse	*Mus musculus*	20	Carp	*Cyprinus carpio*	52
Rhesus monkey	*Macaca mulatta*	21			

Figure 8–2. Haploid Numbers in Different Species

What Are Genes and How Do They Work?

So, we've already established that each chromosome carries the information to build many different proteins, to create traits, and to direct the assembly of a living organism. Each of those proteins' traits and directions was assembled based on the instructions, the sequence of nitrogen bases, in a specific, distinct section of the chromosome. In every organism of that species, the same segment of the same chromosome carries the information for the same trait or protein. There may be different variations of the trait or protein, but they are just different versions of the same thing—like tongue roll or no tongue roll, right thumb or left on top of interlocked fingers, attached or unattached earlobes, or widow's peak or no widow's peak. The segment of the chromosome that codes for a specific protein or for a specific trait is called a **gene**.

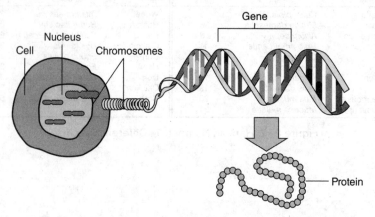

Figure 8–3. How Genes Relate to Chromosomes and Proteins

The Human Genome Project, an international research collaboration that was completed in 2003, established the known sequence and identity of almost all of the genes in the entire human **genome**. The Human Genome Project surprised many scientists when it announced that it estimated the entire human genome, with over three billion nitrogen base pairs, to be made up of a mere 20,000–25,000 genes. This project established that some genes are short, composed of just a few hundred nitrogen bases, while others are long, some containing more than two million nitrogen bases.

BRAIN TICKLERS   Set # 2

Select the correct term to complete each sentence.

1. The Human Genome Project was responsible for determining (<u>how human genes work</u> or <u>what genes are present in humans</u>).

2. Human body cells have nuclei that contain (<u>23</u> or <u>46</u>) chromosomes.

3. The human genome is made up of over (<u>25,000</u> or <u>3,000,000,000</u>) nitrogen base pairs.

(Answers are on page 152.)

What Are Alleles and How Do They Work?

Most eukaryotic species are diploid, which means that each unique chromosome is paired. Each chromosome has the nitrogen base code, or recipe, for all the same proteins as its partner chromosome. The genes for the same proteins are in the same location on each of the paired chromosomes. The paired chromosomes are **homologous** and are called **alleles**. The amazing thing about alleles is that while they code for the same proteins and traits, they can code for different versions of that protein or trait. The two alleles interact with each other in predictable ways to allow an offspring to carry the traits of its parents while being a totally unique individual.

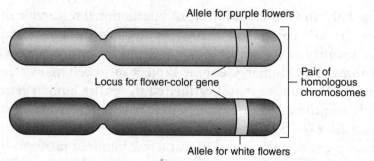

Allele for purple flowers

Locus for flower-color gene

Pair of homologous chromosomes

Allele for white flowers

Figure 8–4. Two Alleles of the Same Chromosome

SUPER BRAIN TICKLERS

Match these definitions to the correct terms.

1. A very specific characteristic of an organism

2. Chromosomes that are homologous are also called _____

3. Gametes, which contain a single chromosome from each chromosome pair, are _____

4. Long strands of DNA that contain the instructions for many proteins

5. For thousands of years, humans have been modifying other species using a process called _____

6. The inheritance of information and traits from one generation to the next

A. haploid

B. selective breeding

C. trait

D. alleles

E. heredity

F. chromosome

(Answers are on page 152.)

Vocabulary

Allele: One of two or more different versions of a gene that are found in the same location on a chromosome.

Chromosome: Long strand of DNA that contains the instructions for many of the traits and proteins needed by all living things.

Diploid: In body cells, chromosomes are paired, with one chromosome having been contributed by each parent at conception.

Gene: A sequence of DNA that provides instructions for a specific protein or trait.

Genetics: The study of heredity, or how traits are passed from generation to generation.

Genome: The known sequence and identity of almost all of the genes in a species.

Haploid: In gametes, chromosomes are singular, so that one chromosome can be contributed by each parent at conception.

Heredity: The inheritance of traits from one generation to the next.

Homologous: A pair of chromosomes each carrying unique information for the same traits, with one contributed by each parent at conception.

Pedigree: A chart that shows family relationships and trait patterns.

Selective breeding: The process of limiting and introducing reproductive opportunities for plant and animal species to other plants or mating partners that have traits that are useful to humans.

Trait: A very specific characteristic of an organism.

Brain Ticklers—The Answers

Set # 1, page 146

1. selective
2. trait
3. DNA

Set # 2, page 149

1. what genes are present in humans
2. 46
3. 3,000,000,000

Super Brain Ticklers

1. C
2. D
3. A
4. F
5. B
6. E

Chapter 9

Mendelian Genetics

Basic Mendelian Inheritance

Have you ever wondered about how you and your siblings or parents look so much alike? In some other families, have you ever wondered why you all look so different?

Gregor Mendel

The history of human understanding of how traits are passed on from generation to generation begins with a man named Gregor Mendel (1822–1884), who was called the "Father of Modern Genetics." Mendel was a monk who lived in a monastery in Austria in the middle to late 1800s. He taught physics, and in his free time he was responsible for maintaining the monastery gardens and growing vegetables to feed himself and the other residents of the monastery. He was also a very alert and thorough investigator.

As he cultivated and grew his gardens, especially between 1856 and 1863, Mendel recognized that his sweet pea plants had very clear traits that appeared to be passed from parent generations to their offspring. He identified their patterns of inheritance. He began to successfully predict the traits of offspring based on the traits of the previous generations. He also kept incredibly detailed handwritten records of these hereditary relationships and traits. The traits that Mendel identified in his sweet pea plants were important because they were unique, they were visible to the naked eye, and each trait had only two variations, or versions. So, Mendel was able to observe whether two parent plants grew to be tall or short, and he kept track all of the later generations for this same trait. Mendel kept track of the following seven distinct and visible traits:

F₁ Crosses for Seven Characteristics In Pea Plants

Characteristics	Dominant Trait ✕ Recessive Trait		F₂ Generation Dominant: Recessive	Ratio
Flower Color	 Purple	 White	705:224	3.15:1
Seed Color and Seed Shape	 Green	 Yellow	6,022:2,001	3.01:1
	 Round	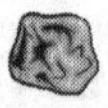 Wrinkled	5,474:1,850	2.96:1
Pod Color and Pod Shape	 Green	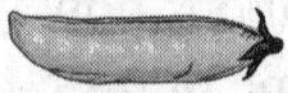 Yellow	428:152	2.82:1
	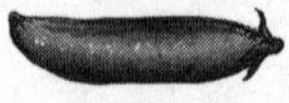 Inflated	 Constricted	882:299	2.95:1
Flower Position and Stem Length	 Axial	 Terminal	651:207	3.14:1
	 Tall	 Dwarf	787:277	2.84:1

Figure 9–1. Sweet Pea Traits Tracked by Gregor Mendel

Mendel used selective breeding to control pollination in his plants. He did this by using a paint brush to delicately remove the pollen from one selected plant and "paint" it onto the pistil of another plant, fertilizing the pea plant ova with specific, known pollen, or sperm. He then carefully documented the traits seen in the many following generations of pea plants. Mendel reported his findings in 1865 and published them in 1866. Although genes and DNA had not yet been discovered, Mendel was able to describe how characteristics are passed from parents to offspring and could accurately estimate the likelihood of a trait's appearance.

PAINLESS FACT

Mendel's discoveries occurred a little over 150 years ago. Can you think of how they are being used today? Could they explain how some members of a family look so similar while others look so different? What about making accurate predictions about the traits of an unborn baby? How might Mendel's work explain the fact that two healthy parents can have both a healthy baby and a baby with a debilitating genetic disease?

BRAIN TICKLERS   Set # 1

Select the correct term to complete each sentence.

1. Gregor Mendel figured out how traits are inherited (<u>before</u> or <u>after</u>) genes and DNA were discovered.

2. Mendel used (<u>green beans</u> or <u>sweet peas</u>) to explore simple patterns of inheritance.

3. Mendel followed the inheritance patterns in his plants for (<u>several</u> or <u>many</u>) generations to better understand how traits were inherited.

(Answers are on page 166.)

Simple Mendelian Inheritance

As we have already discussed, the nucleus of almost every human autosomal (body) cell contains 46 chromosomes in 23 pairs. Each homologous allele (or each member of the pair) contains genes for the same traits, in the same order, as the other allele. **Variation** comes from the fact that traits often come in two or more versions. So, the version of the trait on one allele is often different from the version of the trait carried by the other allele. Mendel's work explored how those two alleles interact with each other.

We now have the technology to understand that the actual structure that carries the code to produce a trait is called a gene and that many genes are assembled, in a very specific order, into each chromosome. Each chromosome then contains genes for different and unique traits. So, for example, one parent may contribute a gene for polydactyly (formation of additional fingers and toes) and the second parent's homologous gene might be for five fingers and toes. Only one of these two possibilities will occur in any children they have, and the likelihood of each variation is genetically predictable. For example, it has been scientifically established that polydactyly, in the absence of other congenital abnormalities, is a **dominant** trait. The gene for five fingers on each hand and five toes on each foot is a **recessive** trait. If one parent has and then passes on (during fertilization) the gene for polydactyly, that child will be born with extra fingers and toes, even though the second parent has the gene for five digits. This happens because the dominant trait masks the effect of the recessive trait when the two are present in the same offspring. The presence of freckles, dimples, a cleft chin, detached earlobes, the ability to roll your tongue, and the left thumb on top of the right when fingers are interlocked are all examples of single-gene dominant traits. Alternatively, having five digits on each hand and foot, smooth hair, and type O blood are recessive traits, so they are only present when the dominant gene is absent.

When we are trying to figure out the likelihood of a trait being passed on from one generation to the next, scientists assign a letter to the trait. That specific letter is not set in stone, but it's a good idea to use a letter that looks different as a capital letter and a lowercase letter. The dominant version of the trait is represented by a capital letter and the recessive trait is represented by the same letter but in lowercase. When we are using letters to represent a variation of a genetic trait we call it a **genotype**, and we always place the capital letter (representing the dominant trait) before the lowercase letter (representing the recessive trait). When both alleles are the same (both either uppercase or lowercase) they are called **homozygous**, and when one allele is upper case (dominant) and the other is lowercase (recessive), the pair of alleles is **heterozygous**. A description of the trait represented by a genotype is called a **phenotype**. Examples of phenotypes would include freckles or no freckles, dimples or no dimples, polydactyly or five digits, and detached or attached earlobes.

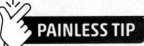

PAINLESS TIP

Here is a table containing each of the possible phenotypes and genotypes in the polydactyly example above, using the letter "F" to represent the dominant trait and "f" for the recessive:

Genotype	Phenotype
FF (homozygous dominant)	2 dominant alleles—polydactyly present (extra digits)
Ff (heterozygous)	1 dominant allele—polydactyly present (extra digits)
ff (homozygous recessive)	2 recessive alleles—polydactyly absent (5 fingers and toes)

Are you surprised that the dominant trait is not the most commonly observed trait in humans?

It's not uncommon! Why do you think that might be?

A hint: We'll cover more about this in the chapter about Evolution and Natural Selection!

With an understanding of Mendelian genetics and of genotypes and phenotypes, we can make statistical predictions about inheritance patterns and the percent likelihood of a trait appearing in the offspring of a specific pair of parents. So far, we have been discussing inheritance patterns for a single gene and a single trait. This is called a **monohybrid** cross. Using this information, and although gametogenesis and meiosis were not yet understood, Gregor Mendel was able to develop the **law of segregation**. This scientific law states that during gametogenesis (the formation of egg and sperm cells) alleles are separated from each other and then are randomly reunited with the same allele from the other parent at fertilization. Mendel's work to understand multiple trait inheritance patterns (two traits are represented in a **dihybrid** cross and three traits form a trihybrid cross) resulted in the development of the **law of independent assortment**. This law states that during gamete formation, each gene is sorted independently of any other genes.

BRAIN TICKLERS   Set # 2

Decide whether each of the following statements is true or false.

1. Recessive traits are rarely seen in human populations.

2. Freckles are a phenotype that might be represented by the genotype FF or Ff.

3. A cleft chin is a genotype that might be represented by the phenotype cc.

4. Human egg and sperm cells contain 23 chromosomes.

(Answers are on page 166.)

Punnett Squares

Punnett squares give us a visual model that helps us track and interpret inheritance patterns and make predictions about the occurrence of traits. We can use Punnett squares to determine ratios and the percent likelihood of parents having offspring with specific traits. Keep in mind that each time offspring are produced, the same ratios or percentages apply. So, unfortunately, if a heterozygous couple has an offspring with a recessive and lethal inherited condition, that doesn't prevent them from having other children with the same disorder.

OK, let's get all the information we can from the following image:

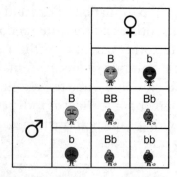

Figure 9–2. A Simple Monohybrid Punnett Square

So, first, we can see that the female parent is represented on the top and she has a genotype of Bb. The male parent is represented on the side and also has a genotype of Bb. We can see that the dominant phenotype is yellow and the recessive phenotype is green. Each parent provides one allele, with one version of the trait in each column or row. For the mother, the columns go from top to bottom, and for the father, the rows go from left to right. The information about the offspring is found in the lower right quarter of the image. The first offspring is homozygous dominant, because both parents gave it a dominant trait (BB), which is yellow. The next offspring, at the bottom of that column, got a B from its mother and a b from its father, making it heterozygous (Bb); it is yellow, because yellow is dominant over green. The mother's second allele (b) is in the next column. Offspring number three got a dominant trait (B) from its father and the mother's recessive (b) trait, making it heterozygous and yellow. The final offspring received a recessive (b) trait from each parent. It is homozygous recessive and green. The genotypic ratio is 1:2:1 (1BB : 2Bb : 1bb) and the phenotypic ratio is 3:1 (3 yellow : 1 green). The genotypic percentages are 25% homozygous dominant / 50% heterozygous / 25% homozygous recessive. The phenotypic percentage is 75% yellow and 25% green. That's a lot of information from one image, isn't it?

Developing a Punnett square for multiple traits can get pretty big. A dihybrid cross Punnett square will have 16 squares for offspring and a trihybrid cross has 64 squares for offspring. There are some rules for completing and organizing these larger Punnett squares. First, always keep the same letter together (like AA or Aa), then place capital letters (representing dominant traits) before lowercase letters (recessive traits), and finally keep paired letters in alphabetical order (like AABbCc). Using these rules will help keep your Punnett squares organized. Accurate Punnett squares can help us make useful predictions about simple and complex inheritance patterns.

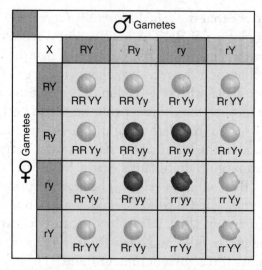

Figure 9–3. Dihybrid Cross Punnett Square

PAINLESS TIP

Remember that each square in a Punnett square will only contain two alleles for each trait, one from Dad and one from Mom. When we fill in the offspring squares, we use one copy of the genotype letter from above that column and one letter from the far left of the row.

OK, weren't you wondering what a trihybrid cross would look like? What information can you get from it?

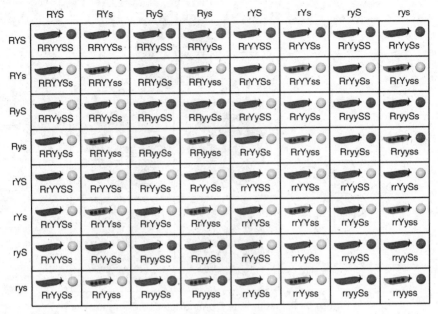

	RYS	RYs	RyS	Rys	rYS	rYs	ryS	rys
RYS	RRYYSS	RRYYSs	RRYySS	RRYySs	RrYYSS	RrYYSs	RrYySS	RrYySs
RYs	RRYYSs	RRYYss	RRYySs	RRYyss	RrYYSs	RrYYss	RrYySs	RrYyss
RyS	RRYySS	RRYySs	RRyySS	RRyySs	RrYySS	RrYySs	RryySS	RryySs
Rys	RRYySs	RRYyss	RRyySs	RRyyss	RrYySs	RrYyss	RryySs	Rryyss
rYS	RrYYSS	RrYYSs	RrYySs	RrYySs	rrYYSS	rrYYSs	rrYySS	rrYySs
rYs	RrYYSs	RrYYss	RrYySs	RrYyss	rrYYSs	rrYYss	rrYySs	rrYyss
ryS	RrYYSs	RrYySs	RryySS	RryySs	rrYySS	rrYySs	rryySS	rryySs
rys	RrYySs	RrYyss	RryySs	Rryyss	rrYySs	rrYyss	rryySs	rryyss

Figure 9–4. Trihybrid Cross Punnett Square

Complex Patterns of Inheritance

Mendel was able to use sweet pea plants to develop and interpret an accurate model for simple inheritance patterns. However, there are other, more complex mechanisms of inheritance for us to understand and enjoy.

Incomplete dominance is a pattern of inheritance that occurs when the appearance of a heterozygous phenotype is an intermediate of the dominant and the recessive phenotypes. A common example of incomplete dominance is a flowering plant called a four o'clock. The dominant phenotype is a red flower, while the recessive phenotype is white. In this species of plant, the heterozygous phenotype is a pink flower.

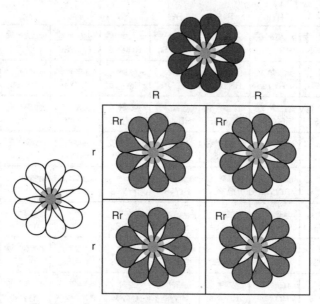

Figure 9–5. Incomplete Dominance F1 Generation

The inheritance pattern called **codominance** indicates that heterozygous alleles are each expressed completely, not as an intermediate phenotype. In the codominance of color genes, the colors often end up appearing speckled. This pattern of inheritance is most often seen in multicolored flowers and in animals like chickens, mice, snakes, and roan horses.

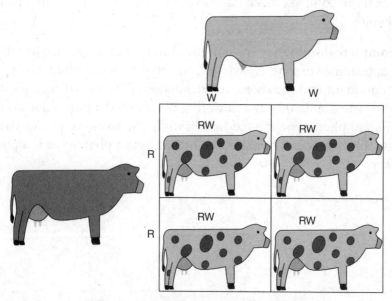

Figure 9–6. Codominance in F1 Generation

BRAIN TICKLERS   Set # 3

Match these descriptions to the correct terms.

1. A black chicken and a white chicken produce speckled chicks

2. 1:2:1

3. Genotypes like aa or bb

4. A blue flower and a red flower produce purple offspring

A. monohybrid cross

B. incomplete dominance

C. codominance

D. recessive

(Answers are on page 166.)

Another important and complex pattern of inheritance is **multiple alleles,** a type of codominance. In this case, though, instead of two there are three or even more possible alleles. One of the most commonly discussed traits governed by multiple alleles is human blood type. You may have heard of people having blood type A, B, O, or AB. This is an example of three alleles dictating blood type; they are represented by the dominant alleles I_A and I_B and the recessive i. These combinations of alleles create blood type A (I_AI_A or I_Ai), type B (I_BI_B or I_Bi), type AB (I_AI_B), or type O (ii).

Polygenic inheritance involves multiple genes in determining traits. Traits governed by multiple genes in humans include height and hair, eye, and skin color. There is a great deal of variation in polygenic traits, and these traits can be influenced by two or more genes. Finally, **epistasis** involves the interactions of different genes where one gene overrides the phenotypic expression of another gene. An example of an epistatic gene interaction would be the action of a gene for albinism, masking the effect of the genes that would otherwise dictate skin, eye, and hair color.

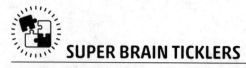

SUPER BRAIN TICKLERS

Match these definitions to the correct terms.

1. The interactions of different genes where one gene overrides the phenotypic expression of another gene

2. Shared inheritance pattern; heterozygous alleles are both expressed completely and not blended together

3. The version of an inherited trait that masks a recessive allele

4. HH, Hh, or hh

5. Polydactyly, freckles, and detached earlobes

6. The blending of dominant and recessive phenotypes

7. Multiple genes working together to produce a spectrum of phenotypes

8. DD and dd

A. dominant

B. phenotype

C. homozygous

D. genotype

E. epistasis

F. polygenic

G. codominant

H. incomplete dominance

(Answers are on page 166.)

Vocabulary

Codominance: Shared inheritance pattern; heterozygous alleles are both expressed completely and not blended together, creating a speckled or spotted phenotype.

Dihybrid: Inheritance pattern for two traits.

Dominant: The version of an inherited trait that will mask the appearance of a recessive trait.

Epistasis: The interactions of different genes where one gene overrides the phenotypic expression of another gene.

Genotypes: A combination of two alleles, which may each be either dominant or recessive.

Heterozygous: Genotype that occurs when the two alleles for a trait carry different versions of the trait.

Homozygous: Genotype that occurs when the two alleles for a trait carry identical versions of the trait. Both can be either the dominant or the recessive trait.

Incomplete dominance: A pattern of inheritance that occurs when the appearance of a heterozygous phenotype is an intermediate of between the dominant and the recessive phenotypes.

Law of independent assortment: Each trait is passed to every offspring independently of any other traits.

Law of segregation: During the formation of egg and sperm cells, alleles are separated from each other and then randomly reunited with the same allele from the other parent at fertilization.

Monohybrid: Inheritance patterns for a single gene and single trait.

Multiple alleles: A type of codominance that involves three or more different alleles.

Phenotype: The appearance, or condition, that results from a genotype.

Polygenic: Inheritance that involves multiple genes in determining traits producing a huge variety of appearances, such as eye, hair, and skin color.

Recessive: The trait that is seen in the absence of the dominant trait. Recessive traits are masked by dominant traits.

Variation: Traits often come in two or more versions; a variation is one of those versions.

Brain Ticklers—The Answers

Set # 1, page 155

1. before

2. sweet peas

3. many

Set # 2, page 158

1. False

2. True

3. False

4. True

Set # 3, page 163

1. C

2. A

3. D

4. B

Super Brain Ticklers

1. E

2. G

3. A

4. D

5. B

6. F

7. H

8. C

Chapter 10

Molecular Genetics

Think about yourself, your family, and your pets (dog, cat, guinea pig, or snake). Consider an elephant at the zoo, an oak tree at the park, and a tomato plant growing in your garden. What about the mushrooms growing so quickly in the grass, or the mold growing around the shower grout in your bathroom? How about the active bacterial cultures growing in the yogurt in your refrigerator? All of these are living things. All of them share all of the characteristics of living things. All of these organisms have DNA that holds the code, or recipe, for them. This DNA doesn't just provide the instructions to make offspring; it also codes for every new cell they need to stay alive. Living cells are always being repaired or replaced. When a Greenland shark dies at 320 years of age (that's right—they live for more than 300 years!), few, if any, of the cells in their bodies were present when they were born. We all have the plans in our DNA to make—hopefully—healthy, new, and exact copies of the cells we were born with. Let's explore how that process works.

A Brief History of the Study of Genetics

First, it seems important to recognize some of the history involved in our current understanding of molecular genetics. This history only goes back around a hundred years. Before that we didn't have the technology to really explore how we inherit traits on a molecular level. In 1928, a microbiologist named Frederick Griffith identified a material that he called the "transforming principle." This material was able to convert a harmless bacterium into one that caused disease. Then, in 1944, a biologist named Oswald Avery and his team developed a way to observe this transformation. Avery and his team continued on to purify and identify the "transforming principle" as DNA.

Other scientists thought, at the time, that transforming material had to have been a protein and that the structure of DNA was much too simple to perform the important and complex job of carrying and protecting hereditary information. In 1952, Alfred Hershey and Martha Chase were able to use viruses tagged with a radioactive phosphorus (remember that phosphorus is found in DNA's sugar-phosphate backbone) to infect bacteria. Their experiments allowed them to confirm that DNA did actually carry the genetic code.

In the early 1950s a number of scientists discovered different aspects of the molecular nature of DNA and how DNA holds and carries the genetic code. Erwin Chargaff was responsible, in 1950, for identifying the adenine/thymine and cytosine/guanine relationships in DNA when his data indicated that adenine and thymine were present in DNA in almost equal amounts and that cytosine and guanine were close to equal as well.

PAINLESS FACT

It may be easier to understand the relationship between adenine/thymine and cytosine/guanine if you review this table, which reflects the DNA base-pairing rules:

Quantity of Nucleotides in DNA				
Source	Adenine	Thymine	Guanine	Cytosine
Calf	1.13	1.11	0.86	0.85
Rat	1.15	1.14	0.86	0.82
Moth	0.84	0.80	1.22	1.33
Virus	1.17	1.12	0.90	0.81
Rat sperm	1.15	1.09	0.89	0.83

Can you see here that adenine and thymine are found in DNA in almost the same percentages, and that cytosine and guanine are also in almost the same percentages?

In 1950, Linus Pauling was able to identify molecular bonding mechanisms that allowed helical structures to form. Pauling's work inspired further discoveries about the chemical structure of DNA by Rosalind Franklin, Maurice Wilkins, James Watson, and Francis Crick. In 1958 Francis Crick coined the phrase "Central Dogma" of molecular biology to explain the sequence of events in copying DNA into an RNA strand and then using this sequence to build a specific protein molecule. While there has been a tremendous amount of new information about DNA and inheritance since then, Crick's central dogma has continued to be the foundation of our understanding of molecular genetics.

BRAIN TICKLERS   Set # 1

Select the correct term to complete each sentence.

1. Avery identified the transforming principle as (<u>protein</u> or <u>DNA</u>).

2. Chargaff discovered that DNA has equal amounts of adenine and (<u>uracil</u> or <u>thymine</u>).

3. The central dogma of molecular biology was established as a (<u>foundation</u> or <u>hypothesis</u>) that explains molecular genetics.

(Answers are on page 181.)

Nucleic Acid

We explored **nucleic acids** a bit in Chapter 3, but let's review. Nucleic acids are one of the four major biomolecule polymers and exist in only two types of macromolecules: DNA and RNA. Nucleic acids are made up of many monomers called **nucleotides**. Nucleic acids are the structures that store our hereditary information and direct the construction of the cells and bodies of all living things. Hereditary information is held in a very specific sequence of the nitrogen bases: adenine (A) bonded with thymine (T) and cytosine (C) bonded with guanine (G). These nitrogen bases are held in their exact sequence by a sugar-phosphate backbone. In <u>D</u>NA the sugar that holds onto the nitrogen base is <u>d</u>eoxyribose, and in <u>R</u>NA it is <u>r</u>ibose.

Three Parts of a Nucleotide

Phosphate

Base (adenine)

Sugar

Figure 10–2. Structure of a Nucleotide

Nucleotides bonded together along the side rails consist of identical sugar-phosphate-sugar-phosphate repeats for the entire strand of DNA, or **chromosome**. The hydrogen bonds that hold nitrogen bases together (A double-bonded with T and C triple-bonded with G) create a double strand that twists into a helix structure. These nitrogen base pairs are the only place in DNA where variation happens. The second strand, which is made up of the paired nitrogen bases to the original strand, is called a **complementary DNA strand**. The order of adenine, thymine, cytosine, and guanine in a strand of DNA gives our cells instructions to build new proteins and cells.

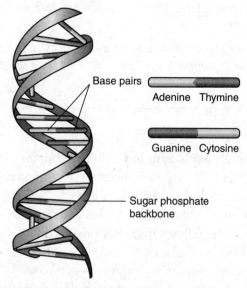

Base pairs
Adenine   Thymine

Guanine   Cytosine

Sugar phosphate backbone

Figure 10–3. Structure of a DNA Molecule

Most DNA is found as chromosomes in the nucleus of eukaryotic cells, with a small amount present in the mitochondria. In prokaryotic cells, DNA is found free-floating in the cytoplasm of bacteria. Each chromosome is made up of shorter coding sections called **genes**. Each gene can be made up of anywhere from 100 to two million nucleotide base pairs. Every gene acts to direct the manufacture of a protein that is necessary in the development, function, activation, or regulation of a healthy organism.

There is one other type of nucleic acid than DNA, and that is ribonucleic acid, or RNA. RNA is single-stranded, uses the sugar ribose instead of deoxyribose, and has the nitrogen base uracil instead of thymine. There are three types of RNA: messenger RNA (mRNA), transfer RNA (tRNA), and ribosomal RNA (rRNA). Each has its own role in taking the code from DNA and building the correct protein structures based on the "recipe" in the sequence of DNA nitrogen bases.

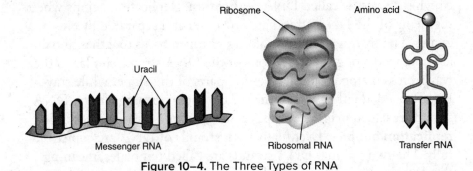

Figure 10–4. The Three Types of RNA

PAINLESS TIP

Sometimes it's easier to study things that can be compared to each other by making a chart or table. Here's one to consider:

DNA:
- Deoxyribonucleic acid
- Deoxyribose sugar
- Double-stranded
- Adenine bonds to thymine
- Found mostly in nucleus

RNA:
- Ribonucleic acid
- Ribose sugar
- Single-stranded
- Adenine bonds to uracil
- Found in nucleus and cytoplasm

BRAIN TICKLERS   Set # 2

Decide whether each of the following statements is true or false.

1. All genes are composed of approximately 500–1,000 nucleotide base pairs.

2. Uracil replaces thymine in RNA.

3. The code in DNA always gives instructions to make proteins.

4. Nucleotides are made of many monomers called nucleic acids.

(Answers are on page 181.)

DNA Replication

During the S phase of the cell cycle, DNA is copied in order for the cell to undergo mitosis. The goal of this type of copying is to produce a pair of cells identical to each other and to their parent cell. This copying process is called **DNA replication**. Replication begins when segments of the DNA strand are unwound and separated by enzymes. The hydrogen bonds holding nitrogen bases together in pairs are unzipped along many sections of the DNA strand, similar to the way a broken zipper opens at central parts of the zipper while staying attached at either side of the break. These openings in the DNA strand are called **origins of replication** and they create spaces called **replication bubbles**. Eventually both strands of DNA are copied fully. The two original DNA strands are called templates, meaning that they both serve as original copies to produce complementary strands. By completing this process, each template strand pairs with a complementary strand to produce a complete DNA molecule to yield two identical sister chromatids to be used in mitosis. This process is called **semiconservative** because one DNA strand is conserved, or kept, while the other side of the DNA strand is new.

REMINDER

Do you remember the study tip from Chapter 3 that "many enzymes end with the suffix "-*ase*"? Let's consider that tip for the enzymes used to perform DNA replication. The double helix of DNA is unwound by the enzyme helicase. Several DNA polymerase enzymes support DNA synthesis. DNA primase prepares surfaces for new bonds to be formed. Topoisomerase creates spaces in the helix to release tension from being wound too tightly and seals those spaces when the copying is complete. Ligase is an enzyme that fills in gaps as replication occurs. Keep watching for other molecules that end with -*ase* for more enzymes.

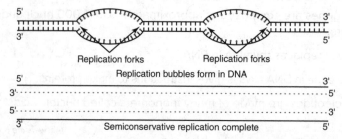

Figure 10–5. Two New DNA Strands Resulting from Replication of Original DNA Strand

So, DNA has more jobs than just replicating to form the DNA for new cells. DNA also directs the formation of proteins by carrying the code for the assembly of proteins. Proteins are important for many reasons. They make up a huge percentage of our bodies, providing a scaffold upon which to build cell, tissue, and organ structures. Actin and myosin are structural proteins that work together to allow muscles to contract and relax. Proteins are needed by living things for growth, development, repair, and maintenance. One maintenance and repair protein is fibrillin, which acts like superglue to maintain cell attachments by holding some of our cells together. Proteins act as metabolic and digestive enzymes (like salivary amylase), as hormones (insulin), and in waste removal and replacement of damaged and aging cells HEXA. If actin and myosin don't work correctly, they can cause muscular and heart problems. When Marfan syndrome occurs, causing fibrillin to be made incorrectly, the layers of the aorta can dissect (separate) or the retina may become detached. When insulin doesn't work correctly, diabetes may result. When HEXA is not made correctly, it causes Tay-Sachs, and as a result lipids begin to accumulate in the brains of babies, ultimately being lethal. Clearly, DNA has a very important role in the accurate production of proteins.

The central dogma of molecular biology is that DNA contains genes that are copied into mRNA during transcription; mRNA leaves the nucleus and engages with a ribosome to be translated to form the very specific protein that is needed at the time. Let's look further into the details of how these processes work.

BRAIN TICKLERS   Set # 3

Match these descriptions to the correct terms.

1. A genetic disorder that causes fibrillin-1 to be made incorrectly. May cause aortic dissection or retinal detachment

2. An enzyme that unwinds and unzips DNA

3. A fatal genetic disorder that causes HEXA to be made incorrectly and lipids collect in the brain

4. Process of copying DNA to produce a pair of cells identical to each other and to their parent cell

A. marfan syndrome

B. tay-Sachs

C. DNA replication

D. helicase

(Answers are on page 181.)

Transcription is the process of copying a gene into the language of RNA. The gene is only a small segment of the DNA strand, so that's the only part that needs to be transcribed. Transcription begins when the cell identifies a need for a new protein to be made; this triggers a transcription complex to be formed from enzymes and other proteins. The transcription complex then finds the beginning of the gene's segment and begins separating the two DNA strands at that location. RNA nucleotides bond temporarily to their partner nitrogen bases. Cytosines and guanines from both DNA and mRNA create these short-term bonds, as do DNA thymines to RNA adenines. One big difference in nitrogen bonding is that DNA strand adenine bases bond temporarily with RNA uracils. Another major difference is that the mRNA, as it begins to separate from the DNA, will remain a single strand and will ultimately leave the nucleus through a nuclear pore. The DNA strand will seal back together and be available the next time this protein needs to be made.

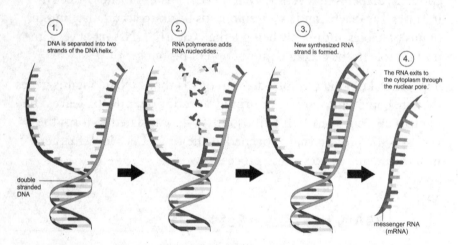

Figure 10–6. The Process of Transcription

So, now that our strand of mRNA has emerged from the nuclear pore into the cytoplasm, **translation** can begin. The mRNA strand moves to a nearby ribosome. Do you remember from Chapter 2 that ribosomes are the organelles that produce proteins? And do you remember that ribosomes are made of protein and rRNA? Here's what that ribosome looks like with a strand of mRNA running through it.

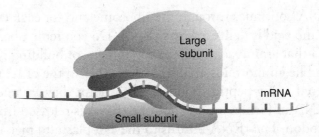

Figure 10–7. Beginning of Transcription—Ribosome and mRNA

As the mRNA travels across the ribosome, its mRNA alphabet of adenine (A), uracil (U), cytosine (C), and guanine (G) is read in groups of three nitrogen bases, called **codons**. As each codon is interpreted at the ribosome, a tRNA is signaled to bring an **anticodon** to match the codon. Transfer RNA, or tRNA, is a molecule with one anticodon on one end and its associated amino acid on the other end. This is a model of tRNA with a codon, an anticodon, and an amino acid.

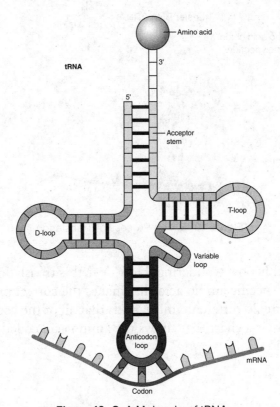

Figure 10–8. A Molecule of tRNA

The mRNA continues through the ribosome and for each codon a new amino acid is delivered by a tRNA. Do you remember from Chapter 3 that amino acids are the monomers, or building blocks, of proteins? The amino acids attach to each other in the order that they are delivered, using peptide bonds. At this point they are released by the tRNA to form a protein and the tRNA is released from the mRNA codon. The tRNA returns to the cytoplasm to pick up its next amino acid. Ultimately, for every three original DNA nitrogen base pairs, there will be one amino acid added to the protein molecule, and the sequence of those amino acids is directed by the gene in the DNA.

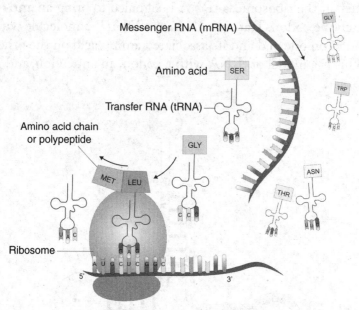

Figure 10–9. The Process of Translation

You're probably now wondering, How does this translation process relate to the specific amino acid that makes the correct protein? Well, there are 20 different amino acids that all of the codons can code for; here is a chart that shows the amino acid coded for by each codon.

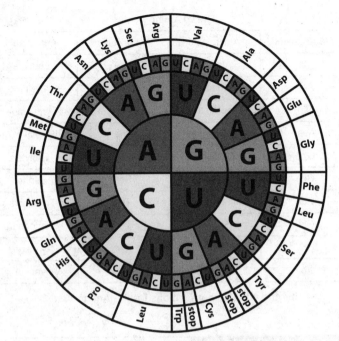

Figure 10–10. Codon Matching Chart

This chart is read by looking at the nitrogen base letter (AGCU) in the center circle first. You pick the first nitrogen base in the codon and select that quadrant of the circle. The second letter of the codon is found in the ring around the center circle. The third letter of the codon is found in the tiniest letters of the next circle that are in line with the first, then second, letters. For example, if the codon was AUG, first go to the A quadrant; at the bottom of the A quadrant is the U section, then in the tiny letters next to the U section find the letter G. The abbreviation next to that third G is Met, which stands for the amino acid methionine.

The 20 Amino Acids

Abbreviation	Amino Acid	Abbreviation	Amino Acid
Ala	Alanine	Leu	Leucine
Arg	Arginine	Lys	Lysine
Asp	Aspartic acid	Met	Methionine
Asn	Asparagine	Phe	Phenylalanine
Cys	Cysteine	Pro	Proline
Gln	Glutamine	Ser	Serine
Glu	Glutamic acid	Thr	Threonine
Gly	Glycine	Trp	Tryptophan
His	Histidine	Tyr	Tyrosine
Ile	Isoleucine	Val	Valine

Figure 10–11. Abbreviations for the Amino Acids

PAINLESS STUDY TIP

It might be a good idea to practice reading the codon chart. What does the codon CCC stand for? How about UUC? Can you read AAA? What about UAC GCA AUU? Try them and you'll find the answers on page 181.

There are codons that code for starting and stopping the protein molecule as well; the codon AUG codes for the amino acid methionine, but it also serves as a code to start the mRNA copying process at the ribosome. The RNA codons UAA, UAG, and UGA code to stop production of the protein.

SUPER BRAIN TICKLERS

Match these definitions to the correct terms.

1. Using a nitrogen base code sequence to assemble the correct molecule of protein

2. A sequential group of three mRNA nitrogen bases that code for a specific amino acid

A. mRNA

B. transcription

C. codon

D. anticodon

3. A segment of a chromosome that codes for the production of a protein

4. A strand of nucleic acid that matches anticodons to codons and brings the correct amino acid to be added to the protein being made

5. The process of copying the DNA sequence of a gene into the language of RNA

6. The cellular organelle that makes proteins

7. A strand of nucleic acid that carries the DNA sequence of a gene to a ribosome to make the correct protein

8. A sequence of three nitrogen bases carried by tRNA that is complementary to the mRNA codon

E. translation

F. tRNA

G. ribosome

H. gene

(Answers are on page 181.)

Vocabulary

Anticodon: A sequence of three nitrogen bases carried by tRNA that is complementary to the mRNA codon. It assures that the correct amino acid will be included in the right place in the protein being assembled.

Chromosome: An entire strand of DNA; a set of individual chromosomes contains instructions for making proteins and for making a new cell or organism.

Codon: A sequential group of three mRNA nitrogen bases that code for a specific amino acid.

Complementary DNA strand: The second and matched strand of DNA, which is made up of paired nitrogen bases to the original strand.

DNA replication: When DNA is copied in order for the cell to undergo mitosis. This type of copying produces a pair of cells identical to each other and to their parent cell.

Gene: A segment of a chromosome that codes for the production of a protein.

Nucleic acid: One of the four major biomolecule polymers; it is made up of only two types of molecules: DNA and RNA.

Nucleotide: A monomer of a nucleic acid; it is made up of a sugar molecule, a phosphate group, and a nitrogen base.

Origin of replication: Openings in the DNA strand where each side begins to serve as a template to form the new, matched strand for the new DNA molecule.

Replication bubble: Gaps in the original DNA strand that open up to allow replication to occur.

Semiconservative replication: When replication is complete, one DNA strand is original while the other side of the DNA molecule is new.

Transcription: The process of copying the DNA sequence of a gene into the language of RNA.

Translation: The process of using the nitrogen base code sequence to assemble the correct molecule of protein.

Brain Ticklers—The Answers

Set # 1, page 169

1. DNA

2. thymine

3. foundation

Set # 2, page 171

1. False

2. True

3. True

4. False

Set # 3, page 173

1. A

2. D

3. B

4. C

Painless Study Tip, page 178

- CCC = Proline
- UUC = Phenylalanine
- AAA = Lysine
- UAC GCA AUU = Tyrosine-Alanine-Isoleusine

Super Brain Ticklers

1. E

2. C

3. H

4. F

5. B

6. G

7. A

8. D

Chapter 11

Genetic Disorders

Basics of Genetic Disorders

Well, now you have lots of information about how DNA and genes carry information about traits and about how we can predict traits in a family using a Punnett square. It's pretty interesting, right? We can take this information to another level, though, by considering how our genes influence our health. It's one thing to consider how genes dictate whether we have blood types A, B, AB, or O, but what about the genes that direct the actual formation of our blood cells? What if their instructions are wrong and cause our red blood cells to form in the shape of a "C" instead of an "O"? This can happen, and it causes a relatively common genetic disorder called sickle cell disease.

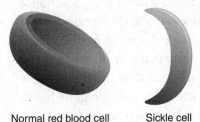

Normal red blood cell   Sickle cell

Figure 11–1. Healthy Red Blood Cell vs. Sickled Red Blood Cell

What if the gene that codes for one of the clotting factors in our blood has an incorrect sequence of codons? Hemophilia is likely to result. An allele for a gene that causes one type of our connective tissues to be made incorrectly can cause a disorder called Marfan syndrome. We'll be exploring these and some of the more than 6,000 other known genetic disorders in this chapter.

There are new genetic diseases and disorders being identified all the time. Our understanding of these disorders is a movement toward developing solutions for them. Approximately 600 of the identified genetic disorders are currently treatable. In the U.S., every 4½ minutes a baby is born with a birth defect. Birth defects affect about 120,000 babies, or one in 33 babies, born in the U.S. each year. Approximately one in 50 people worldwide has a known single-gene disorder. This means that their phenotype, or disorder, results from just two alleles, one from each parent.

Single-gene disorders can be either dominant or recessive. They can result from a parent passing on an existing disease allele for a gene, or it can be a result of spontaneous mutation in a parent's gamete. Another type of genetically influenced disorder is a multifactorial disorder that blends genetic information with infectious processes, environmental factors, and/or aging to produce medical problems like obesity and heart disease. Additionally, chromosomal disorders, which affect approximately one in 263 people worldwide, are a result of chromosomal defects such as extra or missing chromosomes as in Down syndrome and Turner syndrome.

The likelihood of passing on a single-trait genetic disorder follows the same rules as any single-gene trait (remember widow's peaks, tongue rolling, and freckles?). If one of two parents has a gene for a dominant disorder, there will be a 50-50 chance with each pregnancy of the offspring being born with that disorder. If one of two parents has a gene for a recessive disorder, with each pregnancy there is a 50% chance that the offspring will be healthy and a noncarrier of the gene and a 50% chance that the offspring will be an unaffected carrier. If both parents carry a single gene for the disorder; there will be a 25% chance that the offspring will not have or carry the disorder, a 50% chance that the offspring will not be affected significantly but will also carry the gene, and a 25% chance that the offspring will have the disorder.

Dominant Disorders

Let's begin examining some of the genetic disorders caused by a single dominant gene.

PAINLESS TIP

Here is a typical Punnett square for an autosomal dominant inheritance pattern:

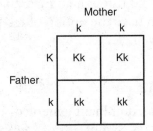

Figure 11–2. Punnett Square for an Autosomal Dominant Disorder (from father)

In this Punnett square, the father has the genotype Kk (heterozygous) and the mother is homozygous recessive (kk). In a dominant disorder, the father will have the disorder because he has one gene for the disorder. The mother does not have the gene or its disorder. The probability, in each pregnancy, that the offspring will receive the gene and have the disorder will be 50 percent.

Huntington's disease

Huntington's disease (HD) is a degenerative brain disorder most often diagnosed in people between the ages of 30 and 50, when symptoms begin to become obvious. It is less commonly seen in children and young adults, and is then called juvenile Huntington's disease (JHD). Symptoms of HD include involuntary tremors and unsteady movements, depression, mood swings, personality changes, forgetfulness, impaired judgement, slurred speech, and trouble swallowing. There is a progression of deterioration through these symptoms over the 10 to 30 years following diagnosis. No cure is available for Huntington's disease, which is ultimately fatal. The dominant gene that causes Huntington's disease is called huntingtin. Everyone has a huntingtin gene on chromosome 4, containing up to 28 CAG sequential nitrogen base repeats. Anyone who has the HD mutation will have 40 or more repeats of the nucleotides CAG, called a CAG repeat expansion, which will cause the person to eventually develop Huntington's disease. Huntington's was one of the earliest genetic disorders that could be identified through genetic testing. Scientists have been able to use predictive genetic testing for the HD mutation since 1986. This testing has presented a dilemma, because many of the people diagnosed with HD already had children before they were aware of their illness. The children are likely to watch their parent go through an incurable, brutal, and ultimately fatal disease and then have the option of testing to find out if that will be their fate as well.

Marfan syndrome

Marfan syndrome is a dominant genetic disorder affecting the body's connective tissue. Symptoms of Marfan syndrome include disproportionate height (Marfan patients tend to be very tall and slender with an arm span longer than their height), flat feet, near-sightedness, crowded teeth, double jointedness, and an indented or protruding sternum (breastbone). The most dangerous effects of Marfan syndrome are aortic dissection, or splitting of the aorta, and retinal detachment. Not all symptoms are seen in all patients, and it's important to keep in mind that these symptoms are also seen in people who do not have Marfan syndrome.

The genetic defect on chromosome 15 that causes Marfan syndrome weakens a type of connective tissue called fibrillin-1. This connective tissue, when healthy, acts like a superglue to hold body parts together. It holds the retina to the back of the eye, it strengthens joints, and it holds and strengthens cardiovascular structures like the aorta and heart valves. Fibrillin-1 in a Marfan patient is much less strong and may allow separations of these tissues, sometimes causing serious (like retinal detachment) or life-threatening (like an aortic dissection) damage to occur.

Approximately one in 5,000 people have Marfan syndrome, and males and females are equally likely to have it. If someone has the gene for Marfan syndrome, he or she will have the disorder. If someone with Marfan syndrome has a child, that child has a 50% chance of also having the disorder. Around 75% of the cases diagnosed with Marfan syndrome have inherited it from a parent. The other 25% of Marfan syndrome patients have it because of a chromosome 15 mutation in an egg or sperm cell at the time of their conception. Once diagnosed, Marfan patients today can live a long, full life. They must be regularly monitored, particularly for cardiac changes, and are likely to require surgeries as needed to support weakened tissues in their bodies. Marfan patients are often discouraged from some athletic (intense contact sports) and recreational (riding roller coasters may trigger retinal detachment) activities that may enhance the risk of a negative outcome. The big picture for Marfan patients though is that with early diagnosis, monitoring, and interventions, and with some modifications in activities, there is a great prognosis for a normal life.

BRAIN TICKLERS   Set # 1

Decide whether each of the following statements is true or false.

1. Approximately 60% of the known genetic disorders are curable.

2. Huntington's disease is usually diagnosed in older adults.

3. Chromosomal disorders happen when both parents have a recessive genetic defect.

4. The child of someone with Marfan syndrome has a 50/50 chance of also having Marfan syndrome.

(Answers are on page 200.)

Recessive Disorders

Next, let's explore some of the genetic disorders caused by recessive genes.

Here is a typical Punnett square for a recessive inheritance pattern:

Parent #1

	C	c
C	CC Homozygous dominant	Cc Heterozygous
c	Cc Heterozygous	cc Homozygous recessive

Parent #2

Figure 11–3. Punnett Square for an Autosomal Recessive Disorder (both parents healthy and heterozygous)

In this inheritance pattern, each parent has one healthy allele and one disease allele. Both parents are heterozygous, and both are carriers of the disorder. With each pregnancy, there is a 25% chance the offspring will receive two healthy alleles and will be healthy. With each pregnancy, there is a 50% chance the offspring will receive one healthy allele and one disease allele; this offspring will be a carrier of the disease and may be healthy or mildly affected by the disease. With each pregnancy, there is a 25% chance the offspring will receive two disease alleles and will have the disease.

Cystic fibrosis

Cystic fibrosis (CF) is one of the most common genetic disorders found in the U.S. Every year cystic fibrosis is diagnosed in one out of every 2,500–3,500 Caucasian American babies. The gene for cystic fibrosis is found on chromosome 7. Mutations can occur at many different locations along the cystic fibrosis gene, also called the cystic fibrosis transmembrane conductance regulator (CFTR) gene. A mutation on the CFTR gene causes the cell to make CFTR protein incorrectly. Abnormal CFTR protein is not able to move chloride (an important ion in the body and a component of NaCl, or salt) to the surface of the cell. Without chloride attracting water to the cell's surface, mucus in the respiratory, digestive, and reproductive systems becomes sticky and thick. Babies with cystic fibrosis develop repeated respiratory issues and salty skin, and they tend to have trouble gaining weight even though they may eat well. Oxygen therapy, medications, feeding tubes, and organ transplants are frequently used treatment options for cystic fibrosis patients.

Hemochromatosis

There are several mechanisms that cause **hemochromatosis,** but the most common is by autosomal recessive inheritance. The frequency of hereditary hemochromatosis (HH) in the U.S. is one case per 200–500 people. The gene that causes HH is found on chromosome 6 and is called HFE. There are several possible sites on this gene that could have a mutation causing different forms of the disease, each form causing iron to accumulate in the body. This "iron over-load" is distributed by the blood—it is toxic and can lead to organ damage. The organs most commonly affected by HH are the liver, heart, and pancreas. Symptoms of HH can be seen in children, but more often begin between the ages of 40 and 60. These symptoms may progress from weakness, fatigue, and joint pain to liver disease, heart issues, **diabetes,** abdominal pain, and skin discoloration. Treatment for HH is generally effective and includes regular removal of blood, treatment of specific symptoms, and dietary changes to avoid foods containing iron.

Sickle cell disease

Another common autosomal recessive disorder is **sickle cell disease (SCD).** This is a group of disorders caused by misshapen red blood cells. There are between 70,000 to 100,000 people with SCD in the U.S. Chromosome number 11 carries a gene that directs the correct development of red blood cells (RBCs). A mutation in this gene causes normally circular RBCs to be shaped like the letter C. These abnormal and rigid blood cells do not carry oxygen efficiently or move smoothly through the capillaries like healthy RBCs. Sickled RBCs become sticky and can block the flow of blood throughout the body. Symptoms include infections, pain, stroke, organ damage, and sometimes death. Treatments are generally specific to symptoms, like pain relief and blood transfusions. Bone marrow and stem cell transplants have been effective in curing SCD.

Tay-Sachs

Tay-Sachs is a rare and incurable disorder most commonly seen and diagnosed in infants around the age of six months. These babies start out in life looking healthy, but the developmental milestones that should start happening around six months often don't happen, and the baby begins to regress. The baby's muscles weaken and acquired skills like rolling over, sitting, and crawling are lost. This disorder happens because a defective HEXA gene on chromosome 15 results in the body not making the HEXA enzyme. Without this enzyme, found primarily in the lysosomes of the brain and spinal cord, certain fats cannot be broken down and disposed of, so they progressively accumulate in the infant's brain. Symptoms include a cherry-red spot in the back of the eye and an exaggerated startle reflex early in the disorder. The progression of Tay-Sachs includes difficulty swallowing, strong muscle twitches, seizures, intellectual decline, and loss of hearing and vision; ultimately, Tay-Sachs is fatal between the ages of three and five. Treatment is simply support and comfort, as there is no cure at this time for Tay-Sachs.

 BRAIN TICKLERS   Set # 2

Match these descriptions to the correct terms.

1. A genetic disorder that causes misshapen red blood cells

2. A recessive disorder that causes thick, sticky mucus to clog up the lungs, pancreas, and reproductive organs

3. A fatal genetic disorder that causes Hex-A to be made incorrectly and lipids to collect in the brain

4. A disorder that causes the blood to carry and deposit excess iron in organs including the liver and heart

A. hemochromatosis

B. tay-sachs

C. sickle cell disease

D. cystic fibrosis

(Answers are on page 200.)

Sex-linked Disorders

Sex-linked disorders are a unique group of usually recessive disorders that have a different inheritance pattern because their gene is found on the X, or female sex, chromosome. Remember, females have two X chromosomes while males have an X and a Y chromosome. The X chromosome is larger and contains around a thousand genes, some of which can cause diseases. The Y chromosome is much smaller and only contains about 70 protein-coding genes, all of which relate to becoming a male and male fertility. As a result, if a female is carrying a gene for a recessive disorder on her X chromosome, the male's Y chromosome will not have a protective allele to prevent that disorder from appearing. One common sex-linked issue is red-green color blindness; here is a Punnett square that shows how it is inherited:

Father (not colorblind)

		X	Y
Mother (carrier)	X	XX female, not colorblind	XY male, not colorblind
	X^c	XX^c female, carrier	X^{c}Y male, colorblind

Figure 11–4. Punnett Square for Inheritance of Red-Green Color Blindness

Hemophilia

One recessive disorder carried on the X chromosome is **hemophilia**, a group of blood clotting disorders. Approximately one out of 5,000 male babies is born with the most common type of this disorder, hemophilia A, which is diagnosed in 400 babies every year. In each type of hemophilia, the reduced numbers of clotting factors in the blood are likely to cause spontaneous or prolonged bleeding, or may cause excessive bleeding following an injury or surgical procedure. The severity of hemophilia is determined by the level of clotting factor present in the blood; the most severe cases are usually diagnosed when the patient is very young. Severe hemophilia is often identified when babies fall as they're becoming more mobile, or when they are teething. In addition to bleeding outside the skin, say from a cut, hemophilia causes internal bleeding. Bleeding into joints can cause chronic and painful joint disease. Bleeding into the head can cause seizures, paralysis, and strokes. Sometimes female carriers of the hemophilia gene can have clotting issues as well. Treatment for hemophilia is usually infusion of clotting factors and, while infusions don't cure hemophilia, they can normalize life for hemophilia patients.

Duchenne's muscular dystrophy

Another sex-linked recessive disorder is **Duchenne's muscular dystrophy (DMD),** a relatively rare muscular disorder. Around the world, DMD affects one in every 3,500 male births each year. The X chromosome carries a gene that regulates the production of a protein called dystrophin, which is important in maintaining cell and tissue membranes in muscle cells. Most children are diagnosed with DMD around the age of three to six. Following diagnosis, DMD progresses from early weakening of pelvic and shoulder muscles to wasting (atrophy) of these muscles. Frequently, DMD patients develop enlarged calves, and most are wheelchair bound in their early teens. Muscles of the torso and arms are eventually involved as the disease progresses. While heart muscle involvement and breathing problems were likely to be fatal to teenage DMD patients in the past, medical advances have allowed many young men with DMD to live into their thirties.

Polygenic and Multifactorial Disorders

Current research has indicated that most diseases and medical conditions have some level of genetic involvement. Polygenic disorders are caused by the interactions between multiple genes, and when the environment and lifestyle influence the interacting genes, these disorders are called complex or multifactorial. The degree to which different genes and the environment interact varies between patients and families. Elements of the environment that can influence genetic disorders include smoking, alcohol consumption, diet, exercise, UV and sun exposure, and exposure to pollutants and carcinogenic (cancer-causing) and mutagenic (mutation-causing) chemicals.

One common multifactorial condition is **heart disease**. A region on chromosome 9 has been associated with an increased risk of heart disease. Heart issues including hypertension (high blood pressure) and atherosclerosis (hardening of the arteries) can run in families and can be minimized or aggravated by individual choices around diet, exercise, smoking, and alcohol consumption. **Type 2 diabetes** and **obesity** are also multifactorial issues, with genes playing a part along with environmental decisions about diet and exercise. Genes on a number of chromosomes (numbers 7, 12, 13, 17, and 20) have been associated with type 2 diabetes. Dietary obesity genes have been found on chromosomes 9, 10, and 15.

Increasingly complex and diverse diseases and disorders include **autoimmune diseases** like multiple sclerosis and lupus; both involve genes on chromosome 6 but may also involve viral exposures, oxidative stress, tobacco smoke, pesticides, and solvents. Genes have been found on chromosomes 1, 14, 19, and 21 that are implicated in inherited forms of **Alzheimer's disease**, a degenerative brain disorder. These genes appear to interact with head injuries and cardiovascular issues. Remember that cardiovascular problems can be aggravated by poor diet, lack of exercise, smoking, and alcohol consumption.

Finally, you may be wondering about the role of genetics and of complex inheritance patterns in **cancer**. Current evidence suggests that around 5–10 percent of all cancers have significant involvement with inherited genetic mutations. Research has also indicated that there are more than 50 hereditary cancer syndromes, which predispose people to cancer development. Cancer is a collection of more than 100 diseases, all caused by cells that no longer follow the rules they should be obeying. Different kinds of cancer are identified by the types of cells, tissues, and organs where the cancer originated. Essentially, cancer is a situation of "cells gone wild." Any of the body's trillions of cells can become cancerous, causing them to grow uncontrolled and then spread (a process called **metastasis**) into other parts of the body. The occurrence of cancer can be a combination of any of these factors: genetic and chromosomal errors, age, family history, many different environmental factors, and often simple random mistakes during cellular reproduction. Today many people survive cancer, often as a result of one or more of these treatments: surgery, chemotherapy, radiation therapy, immunotherapy, targeted therapy, hormone therapy, and bone marrow or stem cell transplant.

BRAIN TICKLERS   Set # 3

Match these descriptions to the correct terms.

1. A degenerative brain disorder that can be a result of multifactorial triggers

2. A sex-linked disorder that causes insufficient clotting factor in the blood

3. A group of diseases caused by uncontrolled cellular reproduction

4. A sex-linked disorder that causes progressive atrophy of skeletal muscles

A. DMD

B. alzheimer's

C. cancer

D. hemophilia

(Answers are on page 200.)

Chromosomal Abnormalities

Chromosomal abnormalities can also cause significant disorders. They are categorized as either numerical, where the number of chromosomes is more or less than 46, or structural, where segments of the chromosome are either missing, added, or rearranged. Most of these abnormalities are caused by problems during meiosis, if pairs of sister chromatids or homologous chromosomes do not separate correctly, a situation called **nondisjunction**. The risk of nondisjunction increases as parents get older. Down syndrome is the result of a numerical chromosome disorder called **trisomy 21**, where the child has an extra chromosome 21. Trisomy 21 occurs in one out of every 800 babies; it causes intellectual disabilities and developmental delays and can be associated with heart defects and thyroid disease.

There are several disorders that are identified with errors in sex chromosome number. **Turner syndrome** occurs with a chromosome count of 45, with one X and no second sex chromosome (genotype XO). This results in a phenotypic female because there is no Y chromosome. The Turner syndrome phenotype involves heart and hearing problems, short stature, a webbed neck, and sterility. When a female has an XXX genotype, she is likely to have reduced fertility and developmental delays. Individuals with **Klinefelter syndrome** have an XXY genotype, are genotypically male, and typically have enlarged breasts, small testes, and reduced body hair.

Genetic Screening

The technology used today for identifying, understanding, and hopefully treating genetic and chromosomal problems is a huge and evolving field. We'll be exploring it in Chapter 12. However, there are some very common and established testing processes that can help identify genetic problems and assist potential and expectant parents who may have concerns about a pregnancy. An **ultrasound** screening is often performed early in a pregnancy to determine a baby's due date and to monitor its early development. Ultrasound can be used along with genetic screening to evaluate risk of Down syndrome and other structural birth defects. An ultrasound test uses a device that emits sound waves across a pregnant woman's abdomen: as the waves bounce off the fetus they are recorded and displayed on a screen, creating an image of the fetus. A clinician will be able to identify many abnormalities in the fetus at this point.

A **chorionic villus sampling** (CVS) is a biopsy of cells taken from the placenta where it attaches to the uterus (chorionic villi cells). This test is usually performed early in a pregnancy, between 11 and 14 weeks. It is generally recommended if there is a family history of inherited genetic or chromosomal abnormalities. **Amniocentesis** is a procedure during which a needle is inserted into the amniotic sac to extract a small sample of amniotic fluid for testing. This procedure is usually performed during the 15th–20th weeks of pregnancy. This procedure is only performed on women who are believed to be at higher risk of having a child with a birth defect. Some of the disorders that amniocentesis can detect include Down syndrome, spina bifida (a neural tube defect), and chromosomal disorders; amniotic fluid can also be DNA tested to look for many genetic disorders, such as cystic fibrosis. A **karyotype** is a test used to identify and evaluate the shape, size, and number of chromosomes in certain body cells. During pregnancy, karyotyping can be done using cells from CVS or amniocentesis to identify chromosomal abnormalities that could affect the baby's development, growth, or body functions.

Down Syndrome — Trisomy 21

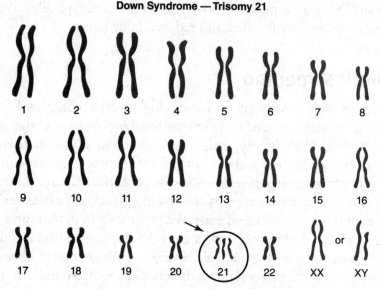

Figure 11–5. Karyotype of a Down Syndrome Patient
(notice three copies of chromosome 21)

Genetic counseling is a service that can provide information and support to individuals or couples who have family histories or risk factors for genetic disorders.

SUPER BRAIN TICKLERS

Match these definitions to the correct terms.

1. Occurs during meiosis, when pairs of sister chromatids or homologous chromosomes do not separate correctly

2. Disorders caused by interactions between multiple genes

3. A procedure during which a needle is inserted into the amniotic sac to extract a small sample of amniotic fluid for testing

4. Genotypical male with an XXY genotype

A. polygenic

B. metastasis

C. ultrasound

D. multifactorial disorder

E. nondisjunction

F. Klinefelter syndrome

5. When cancer cells grow out of control and then spread to other organs

6. When the environment and lifestyle influence polygenic traits

7. A phenotypic female with a chromosome count of 45 with the genotype XO

8. An early, noninvasive screening test often performed to determine a baby's due date and to monitor its early development

G. Turner syndrome

H. amniocentesis

(Answers are on page 200.)

Vocabulary

Alzheimer's disease: A disorder that causes a disruption in brain function.

Amniocentesis: A procedure during which a needle is inserted into the amniotic sac to extract a small sample of amniotic fluid for testing. Some of the disorders that amniocentesis can detect include Down syndrome, neural tube defects, chromosomal disorders, and many other disorders.

Autoimmune diseases: A disorder that causes an individual's immune system to attack its own body.

Cancer: A disease that triggers uncontrolled cellular reproduction, potentially (if treatment is unsuccessful) resulting in spread to other organs. Some cancers are multifactorial.

Chorionic villus sampling: A biopsy of cells taken from the placenta where it attaches to the uterus. It is generally recommended if there is a family history of inherited genetic or chromosomal abnormalities.

Cystic fibrosis: A common recessive genetic disorder that causes the cell membrane to not move chloride effectively, causing respiratory, digestive, and reproductive problems.

Diabetes: A multifactorial disorder characterized by a disruption of the body's ability to metabolize sugar.

Duchenne's muscular dystrophy: A sex-linked disorder characterized by progressive and severe muscular wasting.

Genetic counseling: A service that can provide information and support to individuals or couples who have family histories or risk factors for genetic disorders.

Hereditary heart disease: A multifactorial disorder where genes and environment interact to disrupt heart health and function.

Hemochromatosis: Most commonly presents as an autosomal recessive disorder that causes iron to accumulate in the liver, heart, blood, and pancreas; can be treated by blood removal.

Hemophilia: A sex-linked disorder characterized by the blood's inability to clot.

Huntington's disease: Autosomal dominant neurological disorder, usually diagnosed in late adulthood. Causes tremors with progressive cognitive and motor control decline.

Karyotyping: Testing that can be done using cells from CVS or amniocentesis to identify chromosomal abnormalities.

Klinefelter syndrome: Genotypical male with an XXY genotype; patients typically have enlarged breasts, small testes, and reduced body hair.

Marfan syndrome: Autosomal dominant connective tissue disorder. Causes weakened connective tissue in the skeletal, integumentary, cardiac, and respiratory systems. The biggest risk is aortic dissection.

Metastasis: A process during which cancer cells grow out of control and then spread to other organs.

Nondisjunction: A problem that occurs during meiosis, if pairs of sister chromatids or homologous chromosomes do not separate correctly. Results in gametes either containing the wrong number of chromosomes or chromosomes with rearranged segments.

Sickle cell disease: An autosomal recessive disorder that results in misshapen red blood cells that are sticky and don't carry oxygen efficiently.

Tay-Sachs: An autosomal recessive disorder that results in the accumulation of fats in a baby's brain.

Trisomy 21: An extra chromosome 21 causes Down syndrome. This extra chromosome results in intellectual disabilities and developmental delays and can be associated with heart defects and thyroid disease.

Turner syndrome: A phenotypic female with a chromosome count of 45 with the genotype XO. Phenotype involves heart and hearing problems, short stature, a webbed neck, and sterility.

Type 2 diabetes: A chronic multifactorial disorder in which genes and personal choices about diet and exercise create a disorder which affects the body's ability to regulate and use sugar as a source of energy.

Ultrasound: This screening test is often performed early in a pregnancy to determine a baby's due date and to monitor its early development.

Brain Ticklers—The Answers

Set # 1, page 187

1. False
2. True
3. False
4. True

Set # 2, page 190

1. C
2. D
3. B
4. A

Set # 3, page 194

1. B
2. D
3. C
4. A

Super Brain Ticklers

1. E
2. A
3. H
4. F
5. B
6. D
7. G
8. C

Chapter 12

Genetic Technology

We have explored a lot of different elements of genetics in the last several chapters. We've talked about what chromosomes, genes, and nucleic acids are, how they work, how we can make predictions about phenotypes based on genotypes, and what happens when mutations or dangerous genes are passed on to offspring. We've also covered some early history of gene modification going back to our ancestors selectively breeding crops and animals to be easier to grow, tastier to eat, or just generally more useful to humans.

In the last 50 years, genetic technologies have evolved rapidly in an effort to assist humanity while remaining attentive to an also evolving set of regulations and ethical considerations. Today, we are moving forward into a future with seemingly limitless possibilities for human influence of genetic controls. Where will these alternatives take us and what are our responsibilities to ourselves and to future generations?

Let's explore a few of these technologies.

Polymerase Chain Reaction (PCR)

Polymerase chain reaction, or PCR, is a very commonly used technique that is performed in many laboratories. It can take a very small quantity of DNA and make billions of copies of it in just a few hours, in a process called gene amplification. This chemical reaction takes small amounts of DNA (it can be a sample as small as a single strand of DNA) and exposes it to short DNA sequences, called primers, in order to select the segment of DNA to be amplified. The sample is repeatedly heated and cooled to encourage a DNA replication enzyme to produce many copies of the targeted DNA sequence. PCR allows a very small amount of DNA to be replicated into millions of copies very quickly, allowing scientists to study that DNA in a lab. PCR was crucial in decoding the human genome in the 1990s and is valuable today in many applications, including the diagnosis of genetic disorders, detection of bacteria and viruses, crime scene analysis, paternity testing, and identification of bodies.

DNA Microarrays

DNA chips, or microarrays, first developed and used in 1995, are one of the commonly used technologies in identifying individual genetic profiles and mutations. Microarrays are usually made using a glass or silicon base with labeled DNA attached. The DNA chip is like a computer chip, embedded with short, single-stranded DNA segments, allowing computers to supply, deliver, and read the genetic information that is attached. Microarrays are accurate and quick and can give information about multiple genes using a single test. They are used for disease diagnosis, medical research, and genetic engineering. Diseases that result from genetic mutations, including some cancers, infectious diseases, heart disease, and mental illnesses, are being diagnosed and researched using DNA microchips. Continuing microchip research is likely to allow the evolution of new and individualized medications, gene therapies, and genetically engineered treatments for these disorders.

Another use of microarrays is in individualizing the drug therapy appropriate for every patient. DNA microarrays are crucial to the field of **pharmacogenomics**, where an individual patient's genetic profile is used to define and maximize the effectiveness of his or her treatment plan. DNA microarrays have played a significant role in discovering new genes and in learning how those genes work. Another field that uses microchip science is **toxicogenomics**, a field that explores how an individual's genetic profile will respond to various toxins. This science allows researchers to investigate how toxins affect an individual's cells and the likelihood of offspring inheriting that response.

Home DNA Tests

One of the more common and obvious ways that individuals interact with genetic technology today is by purchasing and using an "at-home DNA test." These tests have you collect saliva (you spit into a tube) or swab the inside of your mouth and then send the sample to the company.

Once received, the company extracts the DNA from the cells in your sample. This DNA is added to a DNA microchip and analyzed to identify unique genetic codes at approximately 600,000 loci (locations) where individual DNA is often found to be different. These loci are called single-nucleotide polymorphisms. DNA home tests can identify whether you carry a gene that might make you sick. It can also tell you if you have a really common version of a gene, or if your version of the gene is less common. These tests can also tell how closely related you are to another member of the company's DNA database, where your ancestors came from, and whether you have some very specific traits.

PAINLESS FACT

Some of the traits that a home DNA test can report include whether you are likely to like cilantro, whether you are afraid of heights, whether you prefer chocolate or vanilla ice cream, whether you are intimidated by public speaking, whether the sound of someone chewing gum annoys you, and what time you are likely to wake up in the morning without an alarm clock. It's amazing how many traits we have that are dictated by our genes, isn't it?

BRAIN TICKLERS   Set # 1

Match these descriptions to the correct terms.

1. A field in which an individual patient's genetic profile is used to define and maximize the effectiveness of his or her medical drug treatment plan

2. A base chip with labeled DNA attached; used in disease diagnosis, research, and treatment and in genetic engineering

A. PCR

B. DNA microarray

C. DNA home test kit

D. pharmacogenomics

3. A lab technique that can take a very small quantity of DNA and make billions of copies of it in just a few hours

4. Individualized genetic tests that can identify whether you carry a gene that might make you sick, where your ancestors came from, and whether you have some very specific traits

(Answers are on page 211.)

mRNA Vaccines

Vaccines made using mRNA to trigger an antibody response to a virus began to get a lot of attention in the media during the COVID-19 pandemic. Vaccinations for COVID-19 became widely available in early 2021. Some people were concerned because they thought it seemed like a new technology. However, mRNA vaccines actually have a more established and longer history that is important to recognize.

First, it helps to understand that the job of vaccines is to teach our immune systems to identify and destroy disease-causing microbes or pathogens. On the surface of pathogens are protein molecules unique to that pathogen, called antigens. Antigens are treated as a threat by our immune systems, which, when encountered, will trigger an immune response in an effort to destroy the invading pathogen. If we've never before been exposed to a particular antigen or pathogen, we can get very sick when we do get infected with it. The job of a vaccination is to expose us to either a dead or weakened pathogen or to fragments of the pathogen (like the proteins or sugars on their surfaces) so that our own immune system will recognize and destroy the invader. Using an mRNA vaccine, we can turn a person's body into its own defender against infection by inducing it to make copies of small sections (the surface spikes) of the COVID-19 virus. The immune system responds by destroying these surface proteins in preparation for future infections. If a vaccinated person encounters this virus again, the immune system is prepared to defend itself. The immune system will then destroy the whole virus as it attacks the spikes that it recognizes as foreign.

The technology behind mRNA vaccines began in1961 with the publication and announcement of the isolation and identification of mRNA. In 1965, structures called liposomes, lipid-based bubbles that could carry materials into cells, were first produced. In 1971, the two (mRNA and liposomes) were unified to produce the first opportunity for the delivery of drugs into a body using liposomes. The delivery of the first vaccine by liposomes occurred in 1974. Many techniques that were developed between 1975 and 2013 enhanced the ability of the pharmaceutical industry to effectively deliver mRNA, or protein-coding abilities. The first human mRNA vaccines against potentially lethal rabies infections began in 2013. By 2015, lipid nanoparticles (LNPs) were being used to deliver vaccines for influenza. When COVID-19 appeared, several pharmaceutical companies were able to move quickly with mRNA vaccination technology that they already had in place to develop vaccines that would help our immune systems identify and destroy COVID-19 protein spikes and, in so doing, destroy the COVID-19 virus particle.

BRAIN TICKLERS   Set # 2

Decide whether each of the following statements is true or false.

1. The job of vaccines is to teach our immune systems to identify and destroy disease-causing microbes or pathogens.

2. If we've never before been exposed to a particular antigen or pathogen, we rarely get very sick when we do get infected with it.

3. Vaccines for COVID-19 were the first time mRNA was used to provide protection from a pathogen.

(Answers are on page 211.)

Genetic Engineering

There are a number of mechanisms to manipulate the DNA of a living thing. All of these mechanisms, collectively, are called genetic engineering. There is an interesting and involved history behind our current understanding and ability to modify genetic code. Some of the genetic technologies commonly used today involve removing unwanted, unproductive, or unhealthy sections of DNA and replacing that DNA segment with a preferred DNA sequence. Let's explore a couple of them.

A **genetically modified organism**, or GMO, has had its DNA altered through genetic engineering. Most often, the modifications made involve replacing original DNA strands with DNA from another organism. When an organism is genetically modified, it is called a **transgenic organism**, and the new DNA it receives could be from another animal, plant, bacterium, or virus. Some common examples of transgenic plant GMOs include crops like tobacco, corn, and soy, which have been modified to improve crop yield and consistency as well as herbicide, fertilizer, and pest resistance. Genetically modified "Golden Rice" has been developed to contain beta-carotene. In humans, beta-carotene is converted into vitamin A, which supports vision, immune systems, and skin. Golden Rice can be used in countries where large segments of the population are vitamin A deficient. In this case, Golden Rice is used mostly to prevent blindness.

PAINLESS FACT

According to the USDA in 2021:

- more than 90% soybean crop contained GMOs
- more than 92% of the corn crop contained GMOs
- more than 90% of cotton crop contained GMOs

In animals, a common genetic modification is the introduction of bioluminescent genes, often from jellyfish or sometimes sea anemones, into fish, rabbits, monkeys, cats, dogs, sheep, and pigs. These transgenic animals are used in research and the fluorescent marker is often attached to a disease-causing gene that is being researched, allowing researchers to easily locate the gene they are working with. Some of the diseases being researched in this way are Parkinson's and Alzheimer's disease, some cancers that result from environmental toxins, and motor neuron diseases. Transgenic technology can also be used to improve farm animal productivity. An example of this is found in the breeding of transgenic goats, whose DNA has had spider DNA inserted. These GMO goats produce milk with silk proteins. Silk protein can then be isolated from the milk to produce a super strong and lightweight silky material that can be used in medical and industrial settings.

These genetic modifications are possible because of the development of a novel gene editing tool called **CRISPR**. CRISPR is an acronym for Clustered Regularly Interspaced Short Palindromic Repeats. Technology using CRISPR can edit genetic code in almost any organism and is cheaper, simpler, and more accurate than any of the previous gene editing techniques. CRISPR is able to cut targeted DNA sequences as directed by a customizable RNA guide. It was discovered in the immune response of bacteria, where CRISPR behaves as a very precise pair of molecular scissors, able to cut specific sections of viral DNA to disable and destroy invading viruses. Once it was discovered that CRISPR was able to be controlled and used as a gene editing tool for bacteria, researchers quickly developed ways to use it as a tool for editing other species genomes. There are two main parts that make up a CRISPR gene editing system: CRISPR-Cas9 nuclease, which binds and cuts DNA like a pair of scissors, and a guide RNA sequence (gRNA), which directs those scissors where to cut. After a segment is deleted, scientists may use CRISPR-Cas9 along with established cellular DNA repair pathways to insert a new or modified gene segment, or to control gene expression. Using current technology, an edited gene will allow the modified organism to be used in research to find diagnostics, treatments, and cures for genetic diseases, as well as having large-scale agricultural and bioenergy impacts.

How does CRISPR-Cas9 work?

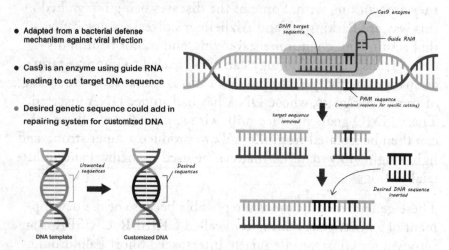

- Adapted from a bacterial defense mechanism against viral infection

- Cas9 is an enzyme using guide RNA leading to cut target DNA sequence

- Desired genetic sequence could add in repairing system for customized DNA

Figure 12–1. CRISPR at Work

Cloning

Cloning is a reproductive process, occurring either naturally or artifi-cially, where identical copies of living things are made. We looked at this phenomenon a few chapters ago when we explored mitosis, a form of asexual reproduction. Scientists can also make identical genetic cop-ies of living things using a process called cloning. Molecules as small as genes can be cloned, as can cells, tissues, organs, and even entire complex animals. Naturally occurring clones include identical twins in humans and other mammals. They happen when a fertilized egg splits early in a pregnancy to produce two embryos with almost identical DNA. These offspring will have the same genetic information as each other but not the same as either parent. Bacteria and some plants also create genetically identical offspring during asexual reproduction. Here, a single parent cell is copied to produce a new, identical offspring.

Artificial cloning has three different types: gene cloning, therapeutic cloning, and reproductive cloning. New copies of genes, or segments of genes, are produced during gene cloning. Gene, or DNA, cloning involves inserting foreign DNA (a gene from one organism) into the DNA of another carrier organism, called a vector. Common vectors include bacteria, viruses, and yeast cells. After the foreign DNA is inserted, the vector is placed into optimal laboratory growing condi-tions, where it reproduces and creates many copies. DNA cloning is commonly used in labs to create genes that researchers want to study.

Therapeutic cloning can be used to develop a line of embryonic stem cells that are genetically identical to an individual. These stem cells can be used in an effort to create replacements for damaged or diseased tissues in that person.

When people refer to cloning, they are usually thinking of reproductive cloning, where a whole new organism is created that is genetically identical to another individual. This type of cloning requires the removal of a mature body cell, like a skin cell, from a donor (the animal that is being cloned). In a laboratory, an egg cell taken from a similar animal is enucleated, meaning that it has its nucleus (with DNA) removed. The nucleus from the donor cell (in this case, a skin cell) is inserted into the enucleated egg cell. The egg cell grows in a test tube for a short time to ensure its viability (its ability to survive) and then it's implanted into the uterus of a mature female of the same species. When the female gives birth, the newborn cloned animal has the same genetic profile as the animal that donated the skin cell. Species of animals that have been reproductively cloned include sheep, cattle, rats, rabbits, cats, deer, dogs, mules, horses, and rhesus monkeys.

 SUPER BRAIN TICKLERS

Match these definitions to the correct terms.

1. An organism with genetic modifications made by replacing original DNA strands with DNA from another organism

2. A gene editing tool that can edit genetic code in almost any organism

3. Can be used to develop a line of embryonic stem cells that are genetically identical to an individual

4. A form of asexual reproduction

5. A GMO grain used to prevent blindness

6. A field that studies how an individual's genetic profile will respond to various environmental chemicals

A. therapeutic cloning

B. toxicogenomics

C. golden rice

D. transgenic organism

E. mitosis

F. CRISPR

(Answers are on page 211.)

Vocabulary

Cloning: A reproductive process, occurring either naturally or artificially, where identical copies of living things are made.

CRISPR: A biological tool for editing the genomes of most living species. It acts like molecular scissors to cut out a segment of DNA and replace it with a preferred DNA sequence.

DNA microarray: A glass or silicon base with labeled DNA attached; used in disease diagnosis, research, and treatment and in genetic engineering.

Genetic engineering: The many different ways to manipulate the DNA of a living thing.

Genetically modified organism: A GMO has had its DNA altered through genetic engineering.

mRNA vaccines: Vaccines that help our immune systems to recognize, identify, and destroy the sugars or proteins present on pathogen surfaces and, in so doing, destroy the pathogen.

Polymerase chain reaction: A technique used in a laboratory that can take a very small quantity of DNA and make billions of copies of it in just a few hours, in a process called gene amplification.

Pharmacogenomics: A field that studies the relationship between an individual patient's genetic profile and the best drug treatment for that patient.

Toxicogenomics: A field that studies how an individual's genetic profile will respond to various toxins.

Transgenic organism: An organism with genetic modifications made by replacing original DNA strands with DNA from another organism.

Brain Ticklers—The Answers

Set # 1, pages 203–204

1. D
2. B
3. A
4. D

Set # 2, page 205

1. True
2. False
3. False

Super Brain Ticklers

1. D
2. F
3. A
4. E
5. C
6. B

Chapter 13

Ecological Relationships

Imagine your life if you lived in a different part of the world. How is your life like that of someone who lives in Sub-Saharan Africa, on a remote Pacific Island, or in the Andes Mountains of South America? What are the factors that influence the similarities and differences? Obviously, there are cultural and language differences, but how much of the culture and diversity is influenced by our physical environments? As human beings, we are an interesting species of animals. We have much more control over our lives and our environments than any other species on Earth, yet we are still impacted by our environment. How do other species of living things interact with the world around them? The study of how living things and their environments impact each other is called **ecology**, and scientists who study this topic are called ecologists.

Relationships Between Species

In order to understand how living things are impacted by their environment, let's first identify that all living things fall into the category of **biotic factors**. Remember that the *bio* in biology means living. So, biotic factors are all of the living elements of an ecosystem, or in the entire world for that matter. Biotic factors also include things that were once living or that are waste products or remains of a living thing. **Abiotic factors** include all nonliving things around us.

 **REMINDER**

Remember, the prefix "a" before a scientific term means *not*; so, abiotic factors are the nonliving elements of our environment.

Biotic factors include members of all the kingdoms of life. They include disease transmission: bacterial infections such as strep, staph, and *E. coli*; Protista infections like malaria; and fungal infections like athlete's foot. Species interactions, like cooperation, **predation**, and **food webs** are also biotic factors. Abiotic factors include all of the chemical and physical elements present in an environment that have an impact on the nearby living species. These include water quality, pH, and availability; soil type and quality; climate (weather patterns, precipitation, temperature); rock types; light availability and intensity; dissolved gases; and natural disasters. As every living thing struggles to stay alive, both biotic and abiotic factors influence their level of success.

Autotrophs and Heterotrophs

There are basically two major kinds of biotic factors. Those that can make food for themselves, the **autotrophs**, and those that can't make their own food, the **heterotrophs**. Remember, most autotrophs make food for themselves by conducting photosynthesis, using energy from sunlight to rearrange the elements in CO_2 and H_2O into sugar ($C_6H_{12}O_6$) and O_2. The products, sugar and oxygen, can then be either used by the autotroph that produced it or be taken in by heterotrophs. Either way, the sugar and oxygen will be used to provide and access energy for cellular activity in the organism that is metabolizing them. Autotrophic organisms that produce sugar and oxygen can also be called **producers**. The organisms that have to use the molecules created by producers are called **consumers**.

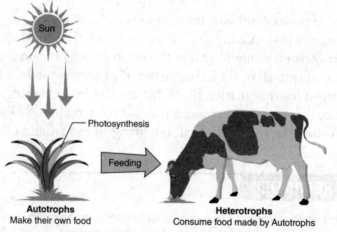

Autotrophs
Make their own food

Heterotrophs
Consume food made by Autotrophs

Figure 13–1. Autotrophs/Producers and Heterotrophs/Consumers

Habitat and Niche

Another important aspect of ecology is recognizing how and where every organism fits into its **ecosystem**. The specific living arrangements for each organism are called its **habitat**. A habitat is the place where an organism makes its home. An example of a habitat would be a grassy field; here you are likely to find many types of grasses, worms, birds, insects, rabbits, and deer, along with bacteria and small fungi in the soil. If you were to visit a beach along the shoreline of the ocean, you might find it to be the habitat of dune grass, sea birds, spiders and insects, sand crabs, and mice. In the study of ecology, it's important to identify the habitats of organisms so that we can understand how to best support those organisms and strengthen the biodiversity of the location.

A **niche** is another important element of an organism's living arrangement. A niche refers to the role an organism plays in its environment. Explaining a niche is more complicated than explaining a habitat because it relates to all the ways an organism impacts all the other organisms in its environment. Let's consider the niche of a rabbit at your local park. How does that rabbit fit into the park's food web, what does it eat, and what eats it? Is the soil chemistry altered by the rabbit's waste? Does the presence of the rabbit's burrow prevent some plants from growing and, in doing so, alter the development of sheltering plants, photosynthetic activity, or food availability? And then, with the rapid reproductive rate of rabbits, there's the question of the effect of large populations of rabbits in a specific location. Understanding the rabbit's niche involves asking all these questions so we can understand best how the rabbit's presence impacts its environment.

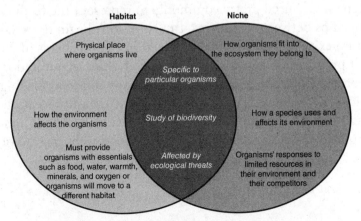

Figure 13–2. Relationship Between Habitat and Niche

PAINLESS STUDY TIP

Venn diagrams can be really helpful when you're trying to compare two or three related topics. They allow you to visualize what aspects are not the same, and then, where the different ovals are merged, their similarities may become clearer. Can you design a Venn diagram for autotrophs and heterotrophs?

BRAIN TICKLERS   Set # 1

Match these descriptions to the correct terms.

1. The role that an organism plays in its environment

2. Autotrophic organisms that produce their own food in the form of sugar and oxygen

3. The place where an organism makes its home

4. The organisms that have to use the sugar and oxygen created by autotrophs to obtain cellular energy

A. producer

B. consumer

C. niche

D. habitat

(Answers are on page 224.)

Feeding Relationships

Food chains are used pretty commonly to represent the feeding relationships between organisms. But they don't begin to explain the complexity of those feeding interactions. Food chains identify a linear progression of who eats whom. But, think about it, do you only eat one thing? In real life, there are many different feeding interactions, with many food sources and many organisms eating these food sources. This variety of interactions means that for accuracy we need to model feeding relationships using food webs instead of food chains.

Let's look at the difference:

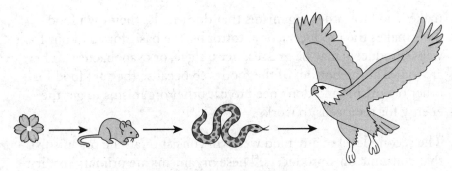

Figure 13–3. A Food Chain

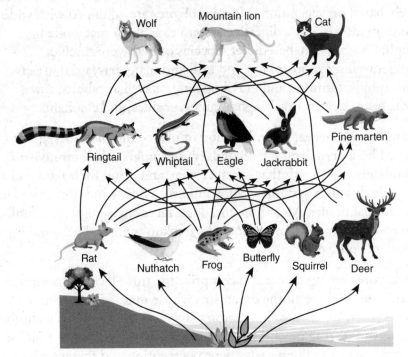

Figure 13–4. A Food Web

Obviously, a food web gives us a whole lot more information about feeding relationships than a food chain. The more information we have, the better we'll understand the ecology of an ecosystem.

Food Webs

Let's explore some of the general relationships between members of any food web.

So, we've already discussed producers, but let's review. A producer is any organism that manufactures its own food, and in doing so also makes food for other organisms that don't make their own food. This makes the producers, or autotrophs, the basis for all of our food webs. Producers may be grasses, trees, algae, or cyanobacteria. They are found at the bottom of the food web because they are food for other organisms and don't need to eat other organisms to get the energy for their cells to work.

The second level of any food web will consist of all the organisms that consume only producers. These organisms are primary (or first-level) consumers called **herbivores**. The diet of an herbivore is exclusively based on vegetation. Most herbivores are equipped with wide grinding molars and a digestive system capable of metabolizing a tough, fibrous plant-based diet. Species of herbivores include water-dwelling zooplankton and krill and many terrestrial insects (like aphids, termites, and grasshoppers), as well as rabbits, cows, koala bears, deer, zebra, hippopotamus, gorillas, and elephants.

Many consumers eat a combination of plant- and meat-based diets. These organisms are secondary consumers called **omnivores**. Omnivores have teeth that can slice, tear, and grind their food and a versatile digestive system equipped to break down and use both plant and animal tissues. Most humans, dogs, raccoons, opossums, skunks, squirrels, chipmunks, mice, sloths, pigs, skunks, many birds, and most bears are omnivores.

Carnivores are predators and comprise the third level of consumers. Carnivores include all the organisms whose main source of nutrition is meat. Carnivores have strong jaws and mostly sharp teeth for killing and tearing meat. They also tend to have eyes in the front of their faces, giving them better depth perception and therefore better hunting ability. Some well-known carnivores include lions, tigers, polar bears, wolves, hyenas, sharks, and honey badgers.

There's a somewhat diverse group of organisms that clean up the waste and remains of living things. They include groups called **scavengers** (also known as **detritivores**), **decomposers**, and **saprotrophs**. Both scavengers and decomposers take care of recycling most organic materials. Animals like worms, insects, crabs, and some birds (like crows, gulls, and vultures) clean up waste in the environment as scavengers. Some organic material is indigestible for scavengers; they can't consume feathers, bones, or fur of dead animals, and detritivores can't digest wood or the cellulose surrounding the cell membranes of plants. The group called decomposers are primarily fungi but also include bacteria, some insects, and snails. They act on dead plant and animal matter by secreting enzymes, breaking the matter down, and then absorbing the small, predigested molecules. Fungi secrete chemicals that decompose organic material, like trees, but can also digest flesh that is still part of a living organism. The final breakdown of any remaining organic waste is taken care of by saprotrophs. Bacteria are the most common saprotrophs that decompose dead animal matter.

BRAIN TICKLERS   Set # 2

Decide whether each of the following statements is true or false.

1. Producers are at the top of the food web because they produce the most food.

2. Carnivores never eat anything other than meat.

3. Scavengers clean up waste in the environment but are unable to digest bones, feathers, or fur.

(Answers are on page 224.)

Species Interactions

There are some organisms that occupy very unique and important positions in different ecosystems. **Apex predators** and **keystone species** are examples of these special members of biological communities. The apex predator in any ecosystem is the top predatory species. In different ecosystems, wolves, eagles, lions, tigers, and great white sharks are the apex predators. When a species is critically important to the survival of an ecosystem, it is called a keystone species. A keystone species' importance in an ecosystem is disproportionately more significant than its population density would suggest. If a keystone species disappears from its environment, other species in that ecosystem are likely to die off. An example of a keystone species is the starfish living along the Washington State coastline. The starfish diet consisted primarily of mussels. This controlled the population of mussels and allowed other species to thrive. When the starfish were removed from the ecosystem during an experiment, the mussel population grew to the point that it crowded out other species. Ecosystem biodiversity was dramatically reduced. This experiment established that recognizing and protecting keystone species can preserve populations of many additional species.

Species interactions are another important biotic factor. There are several different kinds of species interactions that belong to two main categories: **intraspecific** and **interspecific** relationships. Intraspecific relationships are often competitive interactions that occur between members of the same species. Members of the same species may compete for food, shelter, mates, space, light, or dominance. Two oak trees growing next to each other may compete for water, nutrients, and light. The tree that competes most effectively will grow a wider or deeper root system, or it may grow taller than the other. In the end that tree is the most likely to survive longest and produce the most offspring. Another example of an intraspecific relationship is cooperation; we see this in beehives and ant colonies, in which members of the population work cooperatively for the benefit of all. A more extreme form of intraspecific behavior is that of newly dominant male lions in a pride (a group of lions that live together) killing off the cubs of previously dominant males. This intraspecific infanticide ensures that the cubs that the new dominant male helps to raise, train, and protect will carry his genes.

BRAIN TICKLERS   Set # 3

Match these descriptions to the correct terms.

1. The top predatory species in any ecosystem

2. Relationships that occur between members of the same species

3. Perform the final breakdown of any remaining organic waste when the scavengers and decomposers are done

4. A species that is critically important to the survival of an ecosystem

A. apex predator

B. keystone species

C. intraspecific

D. saprotroph

(Answers are on page 224.)

Symbiotic Relationships

Interspecific relationships are interactions between members of different species. Generally, when different species interact, their behaviors are called **symbiotic**, meaning that they have a significant, long-term, close physical association. There are different types of symbioses, including **mutualism**, **commensalism**, **parasitism**, and **predation**. Mutualism is a relationship where both participants benefit from the relationship. An example would be the relationship between a flowering plant and a bee that pollinates it. Both species benefit: the flower gets pollinated and the bee obtains food. Commensalism refers to a relationship where one organism benefits and the other isn't affected. An example of commensalism is the relationship between whales and barnacles. Mature barnacles attach to the skin of whales, where they benefit from water containing nutrients passing them as the whale eats and swims. The whale is not affected by the barnacles. Parasitism is a relationship between organisms with one participant (the parasite) benefiting and the other organism being harmed. It is uncommon for parasites to kill their host because then the host is no longer available to be parasitized. The protozoan parasite that causes malaria in humans is a parasite called *Plasmodium*. When infected, humans are the hosts who are being harmed by the parasite. The *Plasmodium* is moved from host to host by mosquitoes, which act as unaffected intermediate hosts, or **vectors**. Predation is a relationship between species in which one species is a predator (interested in killing food for itself) and the other species is prey (a possible source of food). Wolves hunting moose, penguins catching fish, crows eating the eggs from the nests of other birds, and shrews eating worms and insects are all examples of predator–prey interspecific relationships.

 SUPER BRAIN TICKLERS

Match these definitions to the correct terms.

1. An ecological process in which one organism kills and eats another organism

2. An interspecies relationship where one organism benefits and the other isn't affected

A. vector

B. parasitism

C. predation

D. symbiosis

3. An interspecies relationship where both
participants benefit from the relationship

4. A relationship between organisms with
one participant benefitting and the other
organism being harmed

5. Organisms that act as intermediate hosts for
parasites

6. Organisms of different species that have
significant close physical association and
whose relationships are long term

E. mutualism

F. commensalism

(Answers are on page 224.)

Vocabulary

Abiotic factors: All of the nonliving components of our environment.

Apex predator: The top predatory species in a given ecosystem.

Autotrophs: Organisms that make food for themselves by conducting photosynthesis, using energy from sunlight to rearrange the elements in CO_2 and H_2O into sugar $(C_6H_{12}O_6)$ and O_2.

Biotic factors: All of the living things on Earth, past and present.

Carnivores: Predators and the third level of consumers; includes all the organisms whose main source of nutrition is meat.

Commensalism: An interspecies relationship where one organism benefits and the other isn't affected.

Consumers: Organisms that have to use the molecules (sugar and oxygen) created by producers to obtain cellular energy.

Decomposers: Organisms that secrete enzymes that are responsible for recycling most organic materials.

Detritivores: Organisms, also known as scavengers, that clean up the waste and remains of living things.

Ecology: The study of how living things and their environments impact each other.

Ecosystem: All of the biotic and abiotic elements found in an environment and interacting together.

Food web: A model of all of the feeding relationships in a particular ecosystem.

Habitat: The place where an organism makes its home.

Herbivores: All of the organisms that consume only producers.

Heterotrophs: Autotrophic organisms that produce their own nutrient energy source (sugar and oxygen) by conducting photosynthesis or chemosynthesis.

Interspecific: Relationships between members of different species.

Intraspecific: Relationships between members of the same species.

Keystone species: A species that is critically important to the survival of an ecosystem.

Mutualism: An interspecies relationship where both participants benefit from the relationship.

Niche: The role that an organism plays in its environment.

Omnivores: Secondary consumers that eat a combination of plant- and meat-based diets.

Parasitism: A relationship between organisms with one participant (the parasite) benefiting and the other organism (the host) being harmed.

Predation: An ecological process in which one organism kills and eats another organism.

Producers: Autotrophic organisms that produce their own nutrients in the form of sugar and oxygen.

Saprotrophs: Organisms that decompose any residual organic waste after the scavengers and decomposers are done.

Scavengers: A group of organisms that clean up the waste and remains of living things. They cannot consume feathers, bones, or fur of dead animals.

Symbioses: Organisms of different species that have significant close physical association and whose relationships are long term.

Vectors: Organisms that act as intermediate hosts for parasites.

Brain Ticklers—The Answers

Set # 1, page 216

1. C
2. A
3. D
4. B

Set # 2, page 219

1. False
2. False
3. True

Set # 3, page 220

1. A
2. C
3. D
4. B

Super Brain Ticklers

1. C
2. F
3. E
4. B
5. A
6. D

Chapter 14

Ecology and Planet Earth

How would you describe the environment where you live? Do you have a niche in your environment? What other organisms share your habitat? Do they also share your niche? What abiotic factors influence your environment? Have you ever thought about how geography and weather influence your community or your niche? Have you ever been in a place with a totally different environment from the one you're used to? Does that different climate or landscape alter the way you live your life while you're there? How do you think that different environments might affect the other organisms that live there?

Think about the biotic and abiotic factors that impact your environment. How do you, as a biotic factor, impact your environment?

Levels of Organization

We discussed levels of organization in an earlier chapter. You should remember that atoms bond together to form molecules. Molecules interact to make organelles. Organelles are parts of cells, which are parts of tissues. Tissues, together with other tissues that have similar structures and functions, form organs. Organs together comprise organ systems, and organ systems make up each organism. Now, let's explore the higher ecological levels of organization of the planet Earth.

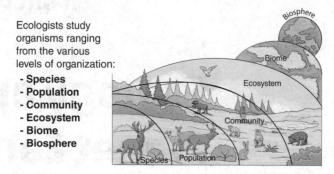

Ecologists study organisms ranging from the various levels of organization:
- Species
- Population
- Community
- Ecosystem
- Biome
- Biosphere

Figure 14–1. Ecological Levels of Organization

So, the ecological levels of organization begin where we left off: with the organismal level of organization. A single organism represents the basis for the level of organization called **species**. A species is generally accepted to be a group of similar living organisms that can breed together to produce fertile offspring. Chimpanzees (animals: mammals), horseshoe crabs (animals: invertebrates), spotted lantern flies (animals: insects), Canada thistle (plants), field mushrooms (fungi), giant kelp (protists), and *Staphylococcus aureus* (bacteria) are all examples of species. *Homo sapiens*, or humans (animals: mammals), are also a species. In taxonomic terms, all identified living things are ultimately classified together with organisms whose bodies have similar structures and functions and that have the same chromosome number, and are called members of the same species.

In biology, though, nothing is absolute. As humans have worked to selectively breed some domesticated species, they have interbred similar species to produce hybrid offspring that have sometimes turned out to be fertile. Mules are useful farm and draft animals—they are the hybrid offspring of a female horse (with 64 chromosomes) and a male donkey (with 62 chromosomes). The mule inherits 63 chromosomes and is almost always infertile because it needs an even number of chromosomes in order to reproduce. Although it is very rare, there have been some reported cases of female mules being fertile, and one case of a male mule producing viable spermatozoa. Another example of human hybridization of two species is that of lions and tigers, different species of big cats from different continents. They would never encounter each other in the wild but have been bred in captivity to produce ligers (male lion bred with female tiger) and tigons (male tiger bred with female lion). This hybridization by humans is generally regarded as unethical, mainly because it's just for "freak show" type exhibition and can result in suffering and unhealthy animals. Also, as climate change has altered the available habitat for polar bears, they have been interbreeding with grizzly bears to produce fertile offspring. So, keep in mind that while species is a term defined as similar animals that can produce fertile offspring, there are some exceptions to that rule, and human behavior is generally the basis for those exceptions.

The next level of ecological organization is **population**. A population consists of a group of organisms of the same species that live together in a specific habitat. Often when we discuss populations, the focus is on the number of individuals present in that group. Different species have different names for their groupings. We are probably most familiar with terms like *herd* for a group of cattle or deer, or perhaps a *pride* of lions, a *flock* of birds, or a *pack* of mules, but many species have their own unique term to describe a population of that species.

PAINLESS FACT

There are some pretty interesting population names for different species. A population of porcupines is called a *prickle*, a group of crows is called a *murder*, a group of locusts is called a *plague*, and a group of sharks is called a *shiver*. Doesn't this make you want to find out the different population names for your favorite species? Doesn't it also make you wonder where some of those names originated (although some seem kind of obvious, like a prickle of porcupines)?

When different populations are present in the same area, they are collectively called a **community**. A community could be the squirrels, chipmunks, robins, cardinals, earthworms, oak trees, maple trees, grasses, mushrooms, and microorganisms living in your neighborhood. Another community might be made up of corals, turtles, phytoplankton, kelp, shrimp, and crabs as well as rays, sharks, eels, and other species of fish. A habitat in Africa might contain populations of grasses, acacia trees, termites and other insects, wildebeests, gazelles, aardvarks, lions, slime molds, and snakes. All of these are interacting groups of living organisms that are adapted to life in their unique habitats.

So, next, we have to consider the unique habitats where these organisms, or biotic factors, live. Their habitats include all of the abiotic factors that make their environments suitable for them to survive. All of the biotic and abiotic factors in a location make up an **ecosystem**. Each of the communities described above will be impacted by their individual terrain (or landscape), water salinity, water depth and availability, soil quality, temperature, access to sunlight, and climate and weather patterns.

PAINLESS TIP

Given what you now know about ecosystems, communities, populations, and species, can you think of all the elements of an ecosystem we haven't yet explored? Maybe one at a shoreline, a lake, or a deep forest that you've visited? Can you assemble all of the biotic factors into a food web? Can you make a Venn diagram of the biotic factors and abiotic factors and see where they intersect?

Biomes

All of the ecosystems in the world belong to **biomes**. A biome is a collection of the ecosystems that exist in similar climatic regions on Earth. Biomes are either water-based or terrestrial. The surface area of the Earth is estimated to be covered by 71% water and is 29% terrestrial, so it makes sense that water biomes account for the largest biomes on Earth. Water biomes are classified based on whether they contain salt water or freshwater. Water biomes include oceans, estuaries, coral reefs, wetlands, and freshwater. Saltwater biomes are inhabited by incredibly diverse populations, including many species of marine algae that produce a large percentage of the Earth's oxygen supply and also take in a great deal of carbon dioxide from the atmosphere.

Oceans are divided into four regions: the intertidal zone (the shoreline, where land and ocean merge), pelagic zone (open ocean), benthic zone (the sandy floor of the pelagic zone), and the abyssal zone (deep ocean, where sunlight doesn't reach). Coral reefs are found mostly in temperate or warm ocean waters around the world and are based on the calcium carbonate remains left by coral populations. Estuaries are where saltwater and freshwater mix as rivers and streams enter the ocean. There are extreme salinity and tidal variations in estuaries, with a high nutrient content and often dense marsh grass and seaweed populations providing a safe place for young ocean life to grow up before entering the pelagic zone where predators await. Wetlands are saltwater or freshwater ecosystems where standing water lasts for at least some of the year. Marshes are wetland systems with a lot of plant and animal diversity. Freshwater biomes include rivers, streams, lakes, and ponds. Salt concentration in freshwater biomes is one percent or less. There are many environmental factors guiding the diversity of **flora** (plant life) and **fauna** (animal life) in a freshwater biome, including water movement, quality, and depth; oxygen availability; and temperature.

Terrestrial biomes include desert, grassland, temperate forest, rain forest taiga, and tundra. A desert biome receives less than 10 inches of rain per year. Temperatures in a desert are associated with its latitude: equatorial deserts like the Sahara and Mojave are almost always hot, polar deserts like the Antarctic and Patagonian deserts are almost always cold, and mid-latitude deserts like the Gobi and Iranian deserts tend to be hot during the day and cold at night. Because of harsh living conditions, biodiversity is low and well adapted to the habitat. Grasslands around the world have deep-rooted grasses as their primary vegetation. Depending on their locations, grasslands are called prairies, savannahs, veldts, pampas, or steppes. Plants and animals in grassland biomes must be highly specialized in order to survive in their specific environments. The African savannah is home to antelopes, elephants, giraffes, and lions, while the North Australian savanna has kangaroos and koala bears.

A temperate forest biome is likely to have four distinct seasons: spring, summer, fall, and winter. The predominant tree in a temperate forest is deciduous; it has broad-leafed trees that shed their leaves in the fall. These trees include oak, maple, elm, chestnut, and hickory. All are trees that form a moderately dense canopy that does allow sunlight through to reach lower growing plants. Mountain laurel and rhododendron grow here, and animal life in the temperate forest includes worms, insects, mice, frogs, rabbits, squirrels, raccoons, opossums, foxes, and deer. A rain forest biome is usually found near the equator, commonly having wet and dry seasons with between 80 and 260 inches of rain per year. Vegetation is layered in the rain forest, with tremendous biodiversity (monkeys, birds, insects, snakes) in the nutrient-rich upper canopy and decreasing biodiversity in the understory, which is closer to the nutrient-poor soil. The rain forest floor is inhabited by ants, earthworms, termites, and fungi that act as decomposers for the huge amount of organic litter that falls from above.

The taiga is a cold northern hemisphere biome and is the largest terrestrial biome on Earth. Taiga is also known as "boreal" forest, meaning that its primary vegetation is evergreen (cone- and needle-bearing) trees. Winter lasts for half the year in a taiga biome and temperatures can drop to as low as −65 degrees F, while summers can be as warm as 70 degrees F. The taiga is a difficult biome to live in, and the plants and animals that live there are well adapted. Spruce, hemlock, and fir trees are the predominant plants and fauna includes worms, insects, trout, salmon, weasels, squirrels, muskrats, foxes, beavers, badgers, moose, and bears. Finally, the tundra is a northern hemisphere biome with very few trees, poor nutrient availability, minimal biodiversity, and a continuously cold climate. Average temperatures in the tundra are around −18 degrees F but can get as cold as −94 degrees F; during the very short summer temperatures can rise to 60 degrees F. A permanently frozen soil layer called **permafrost** exists there all year long between one and three feet below the surface soil. Large plants like trees cannot grow in the tundra because permafrost prevents development of necessary root systems. Flora that does grow in the tundra includes moss, grasses, lichens, and shrubs. Ice melt during the summers triggers marshy tundra, causing the growth and development of thousands of insects (flies, gnats, mosquitoes, and grasshoppers), which become an important source of nutrients for resident and migrating birds (including snow geese, gulls, snowy owls, and arctic loons). Animals of the tundra include fish such as salmon, cod, and trout. Mammals like ermines, squirrels, arctic hares, lemmings, porcupines, wolves, caribou, and polar bears inhabit the tundra.

Finally, the **biosphere** is the biggest and most inclusive level of ecological organization. It is the part of the Earth where all life exists. The biosphere includes all of the biomes, ecosystems, communities, populations, and species. The biosphere reaches from the deepest deep sea vents and throughout the oceans (the hydrosphere), across all land surfaces, the soil and rock (the lithosphere), and up to 30 miles above the Earth's surface (the atmosphere), where some microorganisms are known to exist.

PAINLESS TIP

Isn't it incredible to think about all the places where life can exist? Does it make you wonder how many living species haven't yet been discovered? Or how many became extinct before we even knew they were there? Maybe we'll never know they existed....

What kinds of careers would ask you to investigate these kinds of questions? What training would you need to take on that career path? What kinds of questions do you ask about the world? How would you get more information about those questions and career options?

There are lots of questions here. How do you look for answers to your questions? Many people use the Internet, but what can they do to ensure the reliability of their sources? Here are some suggestions to help you select reputable and credible sources.

Credible Sources:
- Published within the last 10 years
- Peer reviewed by professionals in the subject area
- Written by respected and credentialed authors
- Belong to educational and governmental institutions (URL includes .edu and .gov)
- Presents verifiable data

Non-credible Sources:
- Published more than 10 years ago
- Not peer reviewed
- Written by authors without credentials
- Published on commercial websites (URL includes .com)
- Presents personal opinions

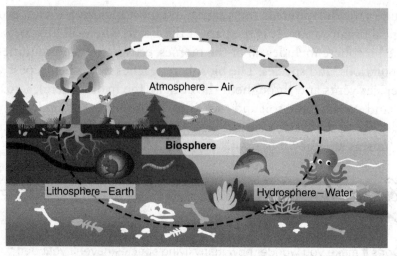

Figure 14–2. Components of the Biosphere

BRAIN TICKLERS   Set # 1

Match these descriptions to the correct terms.

1. A group containing different populations that are present in the same habitat

2. All of the biotic and abiotic factors in a particular location

3. A group of similar living organisms that can breed together to produce fertile offspring

4. A collection of the ecosystems that exist in similar climatic regions on Earth

5. A group of organisms of the same species that live together in a specific area

A. population

B. species

C. ecosystem

D. biome

E. community

(Answers are on page 243.)

Communities and Succession

Understanding how communities and ecosystems change over time is another important element of ecology. A strong and healthy ecosystem develops as a result of the interactions of organisms and populations that live in a stable ecosystem. When changes occur over time in an ecosystem it is called ecological succession. Changes that happen in an ecosystem may be caused by abiotic factors including natural events such as floods, fires, volcanic activity, earthquakes and landslides. Human activity can also trigger ecological succession in the form of clear-cutting forests, building dams, and altering waterways; urban sprawl; and creating a polluted environment. In each of these cases, the existing populations in an ecosystem are likely to be changed by the changes to their environment.

Primary succession occurs in an ecosystem that initially contains no soil. This is often seen after volcanic eruptions; on land, the lava flow covers large areas in solid rock and in the oceans lava flow and tectonic shifts may create new rock-covered islands. In either situation, the terrain is initially inhospitable to plant growth. Without producers there will be no consumers, making this terrain initially uninhabitable. Over time, though, and with the initial growth of lichens, the surface of the rock may begin to crumble to form a thin layer of rocky soil. With some soil cover and seeds blowing in the air, some grasses will begin to grow. Grass roots will continue the job of lichens by breaking up the surface rock more deeply and extensively, producing an increasingly rich soil. These early life forms in uninhabited terrains are called **pioneer species**.

PAINLESS FACT

Lichens are mutualistic, composite organisms made up of algae (or cyanobacteria) and fungi. There are more than 3,600 species of lichens in North America. Lichens can conduct photosynthesis, create nutrients, hold water, and also secrete enzymes to break down organic materials. Lichens are found worldwide in all kinds of ecosystems but are uniquely capable of thriving in the harshest environments, like the polar tundra and volcanic islands.

As more and deeper soil forms, larger and more diverse plants, and then animals, will begin to move in. Then, as they die, their remains increase the richness of the soil, attracting even more diverse species to the area. Over decades and centuries, more diversity develops and finally a stable ecosystem is produced. This ecosystem is called a **climax community**.

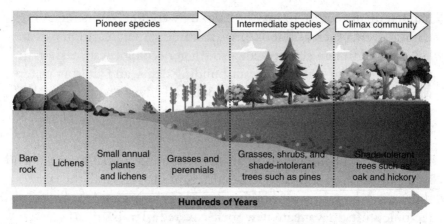

Figure 14–3. Primary Succession

A **secondary succession** is the emergence of a new ecosystem after an existing ecosystem that already contains a soil basis undergoes a catastrophic event that destroys its inhabitants. This often happens as a result of fire, flooding, and landslides. The secondary succession is the progression of populations that ultimately become a climax community. Since the formation of soil isn't an issue, secondary successions usually happen more quickly than primary successions.

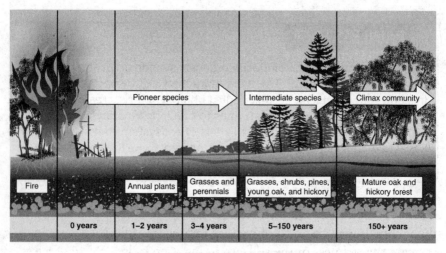

Figure 14–4. Secondary Succession

BRAIN TICKLERS   Set # 2

Match these descriptions to the correct terms.

1. The earliest life forms in uninhabited terrains of primary successions

2. Progression of living occupants in an ecosystem that initially contains no soil

3. The emergence of a new ecosystem after an existing ecosystem that already contains a soil basis undergoes a catastrophic event that destroys its inhabitants

4. A stable ecosystem formed over many years, containing large and diverse populations

A. secondary succession

B. pioneer species

C. climax community

D. primary succession

(Answers are on page 243.)

Ecological Relationships and Energy Flow

As an ecologist explores the relationships of organisms in an ecosystem, it becomes clear to him or her that there are proportional relationships between the roles and total mass of organisms, the numbers of organisms, and the energy exchanged between organisms. These relationships can be modeled by diagramming them into pyramids.

A **pyramid of biomass** tells us that as we move up the pyramid from producers to primary consumers (or herbivores), to secondary consumers (or omnivores), and then to tertiary consumers (or carnivores), the biological mass of the organisms is reduced by approximately 90 percent at each level.

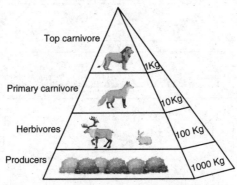

Figure 14–5. Pyramid of Biomass

A **pyramid of numbers** indicates that as we move up through these same levels, there are significantly fewer members of the populations at each higher level.

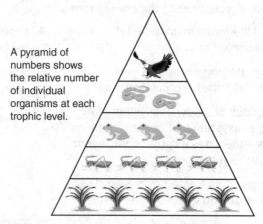

A pyramid of numbers shows the relative number of individual organisms at each trophic level.

Figure 14–6. Pyramid of Numbers

Finally, in a **pyramid of energy**, we can see that as we travel up the successive levels, the energy available to the consumers is reduced by around 90 percent. This means that at each level the consumer only receives 10 percent of the energy that was originally available in that meal; 90 percent of the original energy is lost in the search/hunt for food and in the act of (killing and) eating, digesting, and metabolizing the food source.

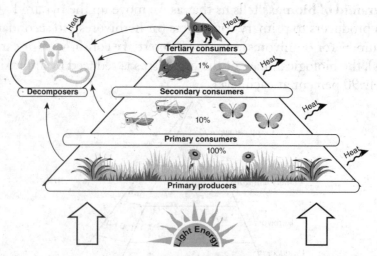

Figure 14–7. Pyramid of Energy

BRAIN TICKLERS   Set # 3

Match these descriptions to the correct terms.

1. A model showing the reduction of nutrient power available as organisms eat their prey

2. A model showing that as we move up the pyramid, the living material of the organisms is reduced by approximately 90 percent at each level

3. A model showing that as we move up through the ecological levels, there are significantly fewer members of the populations at each higher level

A. pyramid of numbers

B. pyramid of energy

C. pyramid of biomass

(Answers are on page 243.)

PAINLESS STUDY TIP

Models can be really helpful when you're trying to visualize relationships between things. Models can be as simple as labeled images, or they can also be as complex as a diagram of all the feeding relationships of organisms in the open ocean. The ecological pyramids may give you a strong sense of how you can better understand the proportions and interactions between different species. Can you think of other ways you might use models? Hint: check out the next section.

Nutrient Cycling in Ecosystems

We explored the atoms and molecules that are essential to life in an earlier chapter (Chapter 3). Let's look, now, at several of those elements as they circulate through ecosystems to become available to living organisms and then are recycled back to be used again and again....

It is important to understand how water moves through the Earth's ecosystems. Do you remember that about 71 percent of the Earth's surface is covered in water and that all living things are mostly made of water? Humans are approximately 60% water, fish are around 80% water, and plants are made of about 80–90 percent water. Exploring the water cycle and its ecological impacts makes sense, and since around 70 percent of the Earth's water is contained in oceans, let's start our investigation there.

When sunlight hits the ocean's surface, it warms the water, resulting in evaporation of water vapor back into the atmosphere. Water also enters the environment through transpiration, which is the release of water through the leaves of plants, as well as organismal cellular respiration and excretion. Water vapor in the atmosphere condenses to form clouds and then, ultimately, returns to the Earth's surface in the form of precipitation. This water is then stored on Earth as ice and fresh, or salt, water. Organisms use these molecules of water temporarily and the water molecules are continuously recycled through the water cycle.

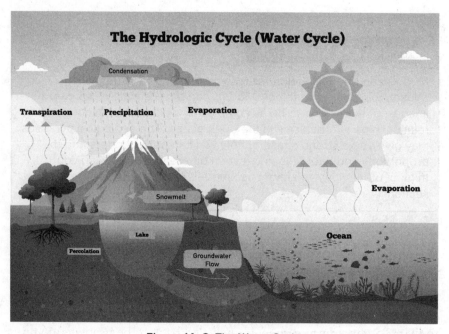

Figure 14–8. The Water Cycle

Carbon is constantly being cycled through the environment during the carbon cycle. Carbon is the core element of all life forms; it is used by plants during photosynthesis and is a byproduct of cellular respiration. Carbon is the element that forms the molecular backbones of carbohydrates, proteins, lipids, and nucleic acids. Carbon is moved through the food web as organisms eat and metabolize their food. As decomposition of dead organisms and organic waste occurs, carbon is cycled through the soil to be picked up by plants and either used by those plants or eaten by consumers to ultimately reenter the cycle as organic waste. Carbon is also found in abiotic sources, including atmospheric carbon dioxide, and as a result of burning fossil fuels.

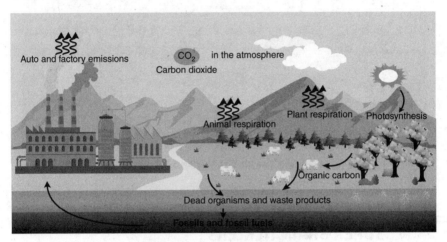

Figure 14–9. The Carbon Cycle

Another essential element is nitrogen. Nitrogen is a critical part of all amino acids and, therefore, of all proteins. In living things, proteins are responsible for scaffolding and structural components, enzymes, transport, hormones, and cellular communication and repair. When any living thing dies, its elements return to the environment. As these bodies and their amino acids and proteins break down, the nitrogen-based molecules that they are made of progress through a sequence of events called decomposition. Bacterial decomposers in the soil convert the amino acids into ammonia, which is turned into ammonium, then into nitrites, and finally into nitrates. Plants can take in nitrates to make proteins. Nitrates can alternatively be denitrified and returned to the atmosphere as simple nitrogen. Atmospheric nitrogen is then cycled back into the soil to reenter the food chain.

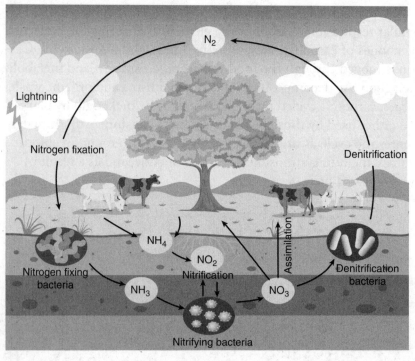

Figure 14–10. The Nitrogen Cycle

SUPER BRAIN TICKLERS

Match these definitions to the correct terms.

1. A collection of the ecosystems that exist in similar climatic regions on Earth

2. Nutrients in H_2O that recirculate through an ecosystem

3. All of the biotic and abiotic factors in a particular location

4. A permanently frozen soil layer that exists all year long in the tundra

5. The parts of the Earth where all life exists

6. A group containing all of the different populations that are present in the same habitat

A. permafrost

B. water cycle

C. community

D. biosphere

E. biome

F. ecosystem

(Answers are on page 243.)

Vocabulary

Biome: A collection of the ecosystems that exist in similar climatic regions on Earth.

Biosphere: The parts of the Earth where all life exists: the hydrosphere, the lithosphere, and the atmosphere.

Climax community: A stable ecosystem formed over many years, containing large and diverse populations.

Community: A group containing different populations that are present in the same habitat.

Ecosystem: All of the biotic and abiotic factors in a particular location.

Fauna: Animal life in an environment.

Flora: Plant life in an environment.

Nutrient cycling: The movement of elements like carbon and nitrogen, and molecules such as water, as they circulate through ecosystems to become available to living organisms and are recycled to be used again and again.

Permafrost: A permanently frozen soil layer that exists all year long; it is between one and three feet below the surface soil.

Pioneer species: Earliest life forms in uninhabited terrains of primary successions.

Population: A group of organisms of the same species that live together in a specific area.

Primary succession: Progression of living occupants in an ecosystem that initially contains no soil, often found on newly formed islands or following volcanic eruptions.

Pyramid of biomass: A model showing that as we move up the pyramid, the living material comprising the organisms is reduced by approximately 90 percent at each level.

Pyramids of energy: A model showing that as we travel up the successive levels, the energy available to the consumers is reduced by around 90 percent, and the rest is lost as heat energy.

Pyramid of numbers: A model showing the reduction of nutrient energy available as organisms eat their prey.

Secondary succession: The emergence of a new ecosystem after an existing ecosystem that already contains a soil basis undergoes a catastrophic event that destroys its inhabitants.

Species: A group of similar living organisms that can breed together to produce fertile offspring.

Brain Ticklers—The Answers

Set # 1, page 232

1. E
2. C
3. B
4. D
5. A

Set # 2, page 235

1. B
2. D
3. A
4. C

Set # 3, page 237

1. B
2. C
3. A

Super Brain Ticklers

1. E
2. B
3. F
4. A
5. D
6. C

Chapter 15

Population Biology and Human Ecology

Can you think of a situation when a population seemed out of balance? A time when there were too many, or too few, organisms in a space together? Can you imagine a new pet gerbil owner that bought a pregnant female and then took too long to identify and separate her male and female pups after they were born? The original gerbil and all her female offspring are likely to become pregnant and have another round of pups in the next few months. The population will rise from one to around seven to approximately 40 gerbils. This population of gerbils, without controls, will be over a hundred in less than a year. What happens when a population overcrowds their space? Let's consider how and why populations change, how different populations and communities can impact each other, and the niche of humans in population biology.

How Populations Grow and Change

So, first let's be sure we remember what a population is. That's right— it's a group of living organisms that belong to the same species living together in the same area. And let's consider what you already know about populations. Even without reading about it here, you know that births and deaths are easy items to consider in understanding population changes. Let's do a little bit more investigation into population change.

First of all, we need to consider that if every organism on Earth were to reproduce without any controls, the Earth would be quickly overrun and would run out of resources. A simplified story of population ecology might say that in order to maintain a perfect population balance, each parent organism would produce one surviving offspring and then die. In that instance, for every organism there is a single replacement organism. But that's not how it works: there are many other factors involved, particularly because of feeding relationships. Consider reproduction in most fish species, an external process called spawning, where females release eggs and males release sperm into their aquatic environment. Huge numbers of eggs are fertilized in this way, but very few of the young fish survive long enough to produce their own offspring. Some die because they are poorly adapted to their environment, but many simply become someone else's meal, a piece of the food web.

Still, there will often be more members of a population with each successive generation. This increase in population occurs because of an increasing birth rate, survival rate, and movement of organisms into that environment (**immigration**). Loss of population members is usually a result of deaths (the **mortality rate**) or because of movement out of an area (a process called **emigration**). Populations without factors significantly controlling their growth rate will usually grow exponentially, and if their numbers are plotted on a graph, it will look like the letter "J," showing exponential growth. A J-curve represents a population's **biotic potential**, the maximum possible growth of a population. It occurs when there is unrestricted growth with the highest possible birth rate and lowest mortality rate. Ultimately, an environment can only sustain an increasing population for so long. At some point, environmental resistance begins to limit population growth. This limitation is called the **carrying capacity**. When a population reaches its carrying capacity in an environment, it begins to stabilize and its graph forms an "S" shape, indicating logistic growth.

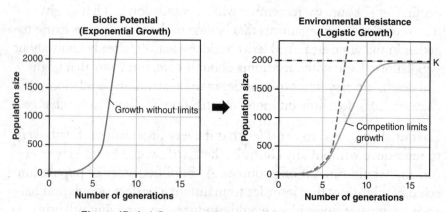

Figure 15–1. J-Curve and S-Curve of Population Growth

REMINDER

Reading and understanding graphs is an important skill for ecologists and for all biologists. Why do we use a table, a line graph, a bar graph, or a pie chart? How do we set them up and interpret them? We have already covered a bit of this, but it's important and bears repeating. So, first, numbers can become a little overwhelming sometimes. A table or graph helps to organize them and eliminates extra words. A chart is used to represent and compare parts of a whole. A bar graph compares categories. A line graph compares independent (x-axis) and dependent (y-axis) variables and shows data that changes continuously over time.

Limits to Growth

Every environment has a carrying capacity. This means that every ecosystem has a limited amount of resources to provide for its inhabitants. These limitations are called **limiting factors**, and may be in the form of water, space, shelter, or food. Limiting factors may also develop as overcrowding intensifies and disease processes or extreme aggressiveness become more commonplace. Limiting factors that result from overcrowding are called **density-dependent limiting factors**, and those that are caused by issues unrelated to population density are called **density-independent limiting factors**.

BRAIN TICKLERS   Set # 1

Select the correct term to complete the sentence.

1. Population changes that are caused by issues unrelated to population density are called (<u>density-dependent</u> or <u>density-independent</u>) limiting factors.

2. Population changes that result from overcrowding are called (<u>density-dependent</u> or <u>density-independent</u>) limiting factors.

3. Population changes that result from abiotic influences are called (<u>density-dependent</u> or <u>density-independent</u>) limiting factors.

(Answers are on page 255.)

Density-dependent limiting factors are issues affecting population growth that are related to population density, such as predation, disease, herbivory, competition, parasitism, and stress. The relationship between predator and prey is known to be a significant density-dependent limiting factor. For example, consider the relationship between the lemming and the stoat. In Greenland, a small furry rodent called a collared lemming is the primary diet of a weasel-like predator called a stoat. Lemmings reproduce quickly and produce many offspring. Stoats reproduce much more slowly and produce fewer offspring. Their predator–prey cycle goes through predictable four-year cycles. In the first three years, there are lower and then growing populations of lemmings. The balance of lemmings and stoats is relatively stable. In the fourth year, the stoat and lemming populations have grown significantly. The stoat population becomes so large that their normal predation on lemmings decimates the lemming population. When their prey population is no longer readily available, the stoat population begins to decline and then crash. The following year the four-year cycle begins again.

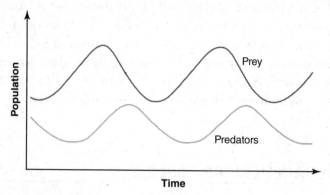

Figure 15–2. Predator–Prey Population Growth Curves

PAINLESS FACT

Not all predators are animals. Venus fly traps and pitcher plants are considered carnivorous plants because they consume insects. Venus fly traps have two hinged leaves that enclose any insect that lands on them, surrounding the insect with digestive enzymes. The pitcher plant catches insects inside a cup, or pitcher, that contains digestive enzymes. These plants usually live in nutrient-poor environments and are able to gain required nutrients as their prey is dissolved and digested.

Herbivores also contribute to population changes. **Herbivory** refers to the interactions between herbivores and the plants they eat.

Herbivores, like deer, behave as predators, and plants are like their prey. When populations of herbivores eat huge amounts of the plants in their ecosystem, the herbivore population may no longer have enough food to eat and both populations are likely to decline.

When populations are crowded, individuals in those populations must compete for necessary environmental resources. Competition is a density-dependent limiting factor because as populations increase, these resources become less and less available to the individuals in that population. When a population reaches carrying capacity, the individuals must begin to compete for scarce resources. Competition in animal communities may be for space, shelter, mating rights, and water and food, while plants are more likely to compete for access to sunlight and soil nutrients.

Disease and parasitism are density-dependent limiting factors because, as populations become more crowded, parasites and disease are more easily spread to other members of the population. Proximity in densely packed populations gives parasites, viruses, bacteria, and fungi more opportunity to move between members of the population through skin-to-skin contact and respiratory or other bodily secretions. Disease-causing microorganisms and parasites must maintain a balance with their hosts because if they become too virulent and begin to kill off members of the host population, the parasites or pathogens are also at risk of dying off.

Finally, stress from overcrowding is another a density-dependent limiting factor. Some species become more aggressive as crowding increases. Stress from overcrowding can induce suppression of an individual organism's immune system. Sometimes, and in some species, stress from overcrowding can trigger females to neglect, kill, or even eat their young. Clearly, density-dependent limiting factors can impact birth and mortality rates as well as the rate of emigration.

Density-independent limiting factors affect population growth and are unrelated to population density. These limiting factors include natural disasters, extreme temperatures, the presence of pollutants, some types of nutrient limitations, and unusual weather patterns. Examples of density-independent factors include the destruction of shallow coral reefs during a hurricane or the decimation of populations following a wildfire or a mudslide. An ongoing severe drought may devastate freshwater fish populations in a lake or river. Most density-dependent limiting factors are biotic (remember, living) and most density-independent limiting factors are abiotic (not living).

BRAIN TICKLERS   Set # 2

Match these descriptions to the correct terms.

1. The maximum growth of a population

2. Something in an environment that causes a decrease in population growth

3. The movement of organisms out of a specific area

4. The number of organisms that an environment can sustain indefinitely

5. The movement of organisms into an environment

A. immigration

B. emigration

C. biotic potential

D. limiting factor

E. carrying capacity

(Answers are on page 255.)

Human Population Growth

Finally, let's explore how human population growth is unique and has worldwide implications.

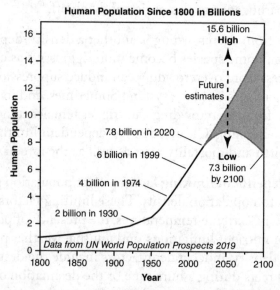

Human Population Since 1800 in Billions

15.6 billion
High

Future estimates

7.8 billion in 2020

6 billion in 1999

Low
7.3 billion
by 2100

4 billion in 1974

2 billion in 1930

Data from UN World Population Prospects 2019

Human Population

Year

Figure 15–3. Human Population Growth from 1800 to 2100

Clearly, the human population on Earth has grown exponentially in the last 200 years. Let's start exploring this by looking back in time. For the humans that lived before the 1800s, limiting factors kept the populations from growing very much or very quickly. Life was harsh and death rates were high. Food was difficult to find, and predators and disease were often deadly. Families often had many children because so many of them were unlikely to survive until adulthood. With the progression of civilization, life became easier for many members of the human race. And the population of humans grew. Throughout the 1800s and the Industrial Revolution, food and materials could be more easily and quickly transported around the world and life became a bit easier and more comfortable. Health care sanitation, nutrition, and medicinal standards were gradually improved and reduced the mortality rates in many areas of the world. Birth rates continued to rise. And the population of humans grew. In 1928, penicillin was discovered by Alexander Fleming. Antibiotics soon became available around the world and significantly reduced the likelihood of death from infection. And the population of humans grew. This exponential growth continued until around 1962–1963, when the human population growth rate reached its peak. The size of the world's population of humans has continued to increase, but the rate of growth has slowed down.

PAINLESS FACT

As of November 2021, the world population was estimated to have exceeded 7.9 billion people. The world population increases by about 83 million people every year. It reached the seven billion mark in October 2011. At the rate we're going, there will be around 10 billion people in the world by 2050.

One element of understanding human population growth patterns is the study of **demography**. Demography is the scientific investigation of human populations. It explores birth rates, mortality rates, economic and social factors, and the age structure of humans in different parts of the world.

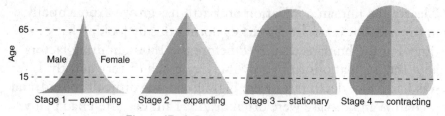

Figure 15–4. Demographic Transition

As scientists have studied demographics, they have found that different countries have progressed from having high birth and death rates through a stage where birth rates remain high while death rates decline. Next, both rates decline and stay low. The final stage involves slight increases in birth rates with continued low death rates. This entire process is called the **demographic transition**. This transition has been seen most rapidly and prominently in the most industrialized countries, like countries in Europe and in Japan and the United States. Another consideration in interpreting and predicting human population growth patterns is a population's age structure. We can expect the least human population growth in countries with the fewest young people and greatest numbers of older adults. We would predict the greatest population growth in countries with the most children, teens, and young adults and with very few older adults.

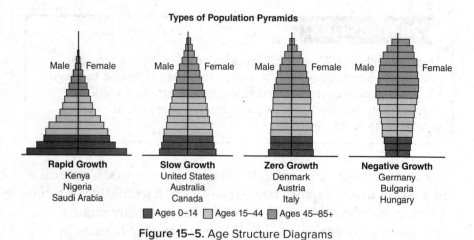

Figure 15–5. Age Structure Diagrams

SUPER BRAIN TICKLERS

Match these definitions to the correct terms.

1. Graph of a population reaching its carrying capacity in an environment and beginning to stabilize

2. Countries with the fewest young people and greatest numbers of older adults

3. The scientific investigation of human populations

4. A group of living organisms that belong to the same species living together in the same area

5. Countries with the most children, teens, and young adults and with very few older adults

6. Graph of a population without factors that significantly control its growth rate and that will usually continue to grow exponentially

A. J-curve

B. S-curve

C. population

D. demography

E. lowest predicted population growth rate

F. highest predicted population growth rate

(Answers are on page 255.)

Vocabulary

Biotic potential: The maximum growth of a population, occurring when there is unrestricted growth with the highest possible birth rate and lowest mortality rate.

Carrying capacity: The number of organisms that an environment can sustain indefinitely.

Demography: The scientific investigation of human populations.

Demographic transition: As scientists have studied demographics, they have found that different countries move between different age structures and high and low birth and death rates. These statistical analyses allow accurate predictions to be made about population growth in different countries.

Density-dependent limiting factor: Issues affecting population growth that are related to population density, such as predation, disease, herbivory, competition, parasitism, and stress.

Density-independent limiting factor: Issues affecting population growth that are unrelated to population density, including natural disasters and unusual weather.

Emigration: The movement of organisms out of a specific area.

Herbivory: The interactions between herbivores (like predators) and the plants (like prey) they eat.

Immigration: The movement of organisms into an environment.

Limiting factor: Something in an environment that causes a decrease in population growth.

Mortality rate: The loss of population members as a result of death.

Brain Ticklers—The Answers

Set # 1, page 247

1. density-independent
2. density-dependent
3. density-independent

Set # 2, page 250

1. C
2. D
3. B
4. E
5. A

Super Brain Ticklers

1. B
2. E
3. D
4. C
5. F
6. A

Chapter 16

Evolution and Natural Selection

If you were asked to name all the living things you could think of, who or what would you name? How many different species could you come up with? You'd likely begin by listing common mammals we interact with, such as humans, dogs, cats, cows, horses, goats, and ferrets. Given a bit more time, maybe you'd think of mosquitoes, flies, fish, snakes, and worms. Would you need some prompting to list flowers, grass, dandelions, trees, and moss? When you think some more, you'd probably remember that animals and plants aren't the only living things. Maybe you'd name mushrooms and bread mold, algae and amoebas, and all the different kinds of bacteria. Would you consider all the living things that once lived on Earth but are now **extinct**? How are all these organisms related to each other? How is a starfish related to a goldfish? What is the connection between a mouse and a giant oak tree, a shark and an amoeba, or a banana and an *E. coli* bacterium? What is *your* relationship with all of these?

We would probably agree that all of these organisms are quite unique and distinctive, but we'd also have to agree that as living things, all of them must share certain basic characteristics. They all grow in size and develop as they age; all of them reproduce, are organized, and maintain homeostasis. All of them adapt to their environment and metabolize nutrients to get rid of waste. Every living thing is assembled based upon instructions encoded in their DNA. The DNA in all organisms is made of the same nitrogen base molecules coding in the same way for the same amino acids and their sequence of assembly. Each species is different simply because the unique proteins being manufactured using these amino acids are assembled according to the instructions for their species, handed down from generation to generation. We can trace human DNA back through time and through many different ancestral mammalian **primates**. When we get back to early primates, we will ultimately encounter a primate ancestor that is also an ancestor of today's apes. That common ancestor is where our history and our DNA intersect with other apes. The great apes developed separately from that ancient common ancestor to become humans, gorillas, chimpanzees, and bonobos.

But that's just a small piece of the story. If we go back even further in time, we'll find DNA connections with less and less similar species. Even further back in time, our DNA connects with that of all other kingdoms. That's right—we are related to all animals, plants, fungi, protists, and even bacteria. Studying these connections and relationships helps us to understand the science of **evolution**.

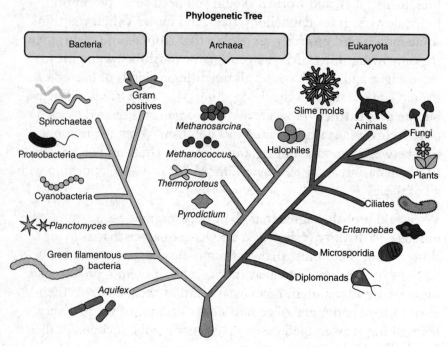

Figure 16–1. Cladogram of Life on Earth

PAINLESS TIP

A **cladogram** is a helpful tree-shaped diagram that shows the evolutionary relationships between living things based, these days, on both physical traits and DNA. The root, or trunk, indicates the initial ancestor, a species that all of the branches have in common.

BRAIN TICKLERS   Set # 1

Select the correct term to complete the sentence.

1. Organisms that once lived on Earth but no longer exist are called (<u>endangered</u> or <u>extinct</u>).

2. The process of organisms changing over time to better adapt to their environment is called (<u>extinction</u> or <u>evolution</u>).

3. A linear diagram that shows how species relate to each other over time is called a (<u>line graph</u> or <u>cladogram</u>).

(Answers are on page 274.)

Evidence of Life on Earth

Scientists have determined that the Earth was formed approximately 4.6 billion years ago. The Earth's atmosphere was extreme and uninhabitable; it was hot, violent, and filled with toxic gases including ammonia and methane. Over millions of years, land masses shifted, merged, and formed continental plates; volcanoes were extremely active; and weather patterns were turbulent. The expansive world oceans churned, full of chemical elements including carbon, hydrogen, oxygen, nitrogen, phosphorus, and sulfur (remember CHONPS?). These elements were mixed, heated and cooled, struck by lightning, and jolted by volcanic eruptions, creating a chemical slurry that scientists call a **primordial soup**. This primordial soup contained molecules that would ultimately become available to form the basic molecules of life. For over three billion years, the terrestrial Earth remained uninhabitable as the primordial soup churned, producing the basic molecules to build the simplest early life forms.

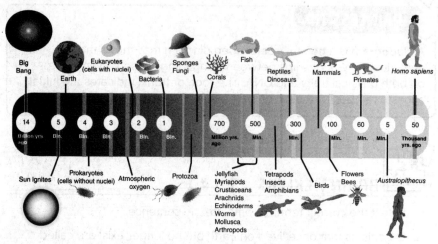

Figure 16–2. History of Life on Earth

The earliest evidence of life, in the form of primitive pre-cells, has been estimated to have appeared approximately 3.8 billion years ago. Simple proteins, lipids, carbohydrates, and nucleic acids interacted, combined, and colonized to create the earliest true cells, bacteria, around 3.5 billion years ago. As these prokaryotes thrived in prehistoric oceans, they evolved and diversified; some gained the ability to use energy from sunlight to synthesize sugars as nutrients. These organisms were the earliest cyanobacteria. As early bacteria evolved, colonized, and became more complex, they began surrounding their DNA with a nuclear membrane. These were the first eukaryotes. Over the course of around 1.5 billion years, ocean dwelling photosynthetic cyanobacteria and blue-green algae were nourishing themselves with the sugars they made during photosynthesis and were releasing oxygen (O_2) as a byproduct. This oxygen began accumulating in the atmosphere about 1.8 billion years ago.

PAINLESS FACT

If you do the math, you'll recognize that the last paragraph describes life forms on Earth for over two billion years! That means two billion years of DNA being passed from generation to generation of living organisms. Mutations and adaptations occurred, allowing many new species and variations of the earliest life forms to develop. Life on Earth was evolving. These organisms are our earliest ancestors!

The last billion years on Earth have been a time of incredible diversification and growth. Around 650 million years ago, the earliest multicellular organisms began to thrive in the world's oceans. The Earth's atmosphere was continuing to accumulate oxygen. Life on this planet was continuing to evolve. By 550 million years ago, the seas were full of life; both prokaryotic and eukaryotic organisms were thriving, and the earliest lichens were establishing themselves on land. Around 460 million years ago, green and red algae, worms, corals, marine invertebrates, and primitive fish were populating the oceans as the earliest vascular plants began to thrive on land. By 370 million years ago, many species of animals as well as seed-bearing plants were living on land. Four-legged vertebrates and land-dwelling arthropods were colonizing terrestrial Earth. Keep in mind that all of these emerging life forms descended from those earliest bacterial cells churning around in the primordial seas.

The largest mass extinction in Earth's history occurred around 250 million years ago. While we can't be absolutely sure of what happened, it has been scientifically accepted that vast numbers of species became extinct at this time. An estimated 96 percent of marine life and 70 percent of terrestrial life were destroyed as a result of this event. After this mass extinction, new and different species had the opportunity to establish themselves. Reptiles began to thrive and initiated the "Age of the Dinosaur." The earliest mammals were tiny mouse-like burrowing animals that appeared and shared the Earth with dinosaurs around 240 million years ago. The earliest flowering plants established themselves around 145 million years ago.

Around 65 million years ago, another mass extinction event occurred, resulting in the extinction of the dinosaurs and many other types of life on Earth. Many life forms did survive this event, though, and they continued to adapt and evolve, repopulating with new and even better adapted species. Approximately 30 million years ago, the environment on Earth was right for a population explosion land mammals and flowering plants. All of these life forms shared DNA with those earliest bacteria, and all continued to adapt and evolve. About 1.8 million years ago, five-fingered primates that were the earliest recognizable ancestors of the human race appeared. They belonged to the genus *Homo*. These primates are the ancestors of lemurs, lorises, tarsiers, New World monkeys, Old World monkeys, orangutans, gorillas, chimpanzees, bonobos, and humans. It is important to note that while we share common ancestors, we do not descend from each other. Humans, or *Homo sapiens*, first appeared in Africa about 60,000 years ago.

Charles Darwin

Charles Darwin (1809–1882) wrote a book called *On the Origin of Species*. In this book, published in 1859, Darwin described the process by which populations of living organisms evolve. The book consolidated Darwin's lifetime observations of the natural world. As a child Darwin loved nature and preferred exploring his environment to classroom learning. In 1831, unsure of the path he was going to take in life, Darwin signed on to travel on the ship *HMS Beagle*. He joined as a cartographer, or map maker, and as a companion to the ship's captain. Their voyage around the world was expected to take two years but turned out to last for five years. When it was over, Darwin had amassed volumes of notes and biological drawings. He had also collected many biological specimens.

When the journey was over and Darwin was back home in England, he settled in to investigate what he had seen and learned in his travels. He spent most of the next 25 years developing his explanation of the relationships between the many organisms he had encountered. One type of animal that drew his attention was birds. He had seen many different types of birds while the *HMS Beagle* traveled through the Galapagos Islands. These islands are an archipelago of more than twelve volcanic islands in the Pacific Ocean about 600 miles west of the coast of Ecuador. Each island formed and developed uniquely from the others; some remained rocky with sparse plant growth while others became lush with tropical plant growth. Darwin ultimately recognized that all these different types of birds were different versions, or species, of finch. He found that the finches were very different on each island. He realized that thriving on each island required unique characteristics. The birds were different sizes and colors, but, more specifically, their beak shapes were appropriate to the type of food that was available on each island.

Adaptive Radiation

Figure 16–3. Galapagos Finches

Darwin identified similar distinctions between the tortoises found on the different islands as well.

The ranges of variations within a species based upon living in different environments is called **adaptive radiation**. These relationships kept on appearing to Darwin as he continued to examine and explore samples and notes from his world travels. Darwin began to recognize that he was seeing organisms that were different from their ancestors. He realized that those differences were the result of adaptations that helped organisms and their offspring to best survive and reproduce in their environments.

BRAIN TICKLERS   Set # 2

Match these descriptions to the correct terms.

1. The diversity of variations within a species based upon living in different environments

2. The earliest recognizable mammalian ancestors of the human race; they still exist today

3. An archipelago of islands in the Pacific Ocean where Darwin identified evidence of natural selection

A. primordial soup

B. Galapagos Islands

C. adaptive radiation

D. primate

4. The mixture of elements that existed in the oceans of ancient Earth, eventually forming molecules that were responsible for the development of life on Earth

(Answers are on page 274.)

Theory of Natural Selection

Charles Darwin was important as a scientist because he was an explorer, a naturalist, and an open-minded thinker of the 1800s. His most important contribution to science, though, was his explanation and recognition of the mechanism by which species evolve: his **theory of natural selection**. Remember, a theory is a scientific statement that is so thoroughly tested and supported that scientists accept it as a basis for understanding and explaining natural phenomena.

Darwin's theory of natural selection can be broken down to five important foundational statements. First, organisms overproduce offspring, but many of these offspring do not survive long enough to reproduce. Second, members of a species are different from each other, and these variations are passed down from parent to offspring. Third, some organisms have variations that make them better adapted to survive in their environment than others. Fourth, better adapted organisms are likely to live longer and produce more surviving offspring than less adapted organisms. Finally, as beneficial adaptations keep an increasing number of offspring alive, those variations become characteristic to the species.

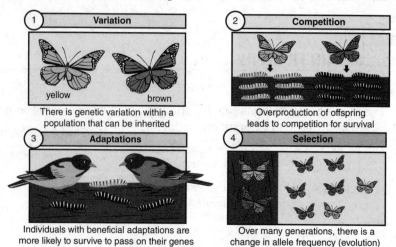

Figure 16–4. Theory of Natural Selection Example

Between 1855 and 1865, while Charles Darwin was working to develop, explain, and publish his theory of natural selection, Gregor Mendel was working in his monastery gardens in Austria and developing his understanding of the inheritance of traits. Mendel's work was published in 1865. Neither Darwin nor Mendel was alive to see how their discoveries came together and enlightened scientific knowledge. In the early 1900s, scientists began to explore the chemical basis of inheritance. They soon began to recognize that, together, Mendel and Darwin had explained the basics of how life has evolved on Earth.

PAINLESS TIP

Remember, a scientific theory is always supported by substantial evidence. Life has existed on Earth for billions of years, but we as individuals only live here for a tiny part of that time. Shouldn't it be difficult to get evidence of populations changing over time? Staying open minded, asking questions, and keeping our eyes open for evidence, like Darwin and Mendel did, is the key to understanding events that happen in distant times and places.

Evidence of Evolution

Darwin's theory of evolution by natural selection is a scientifically accepted explanation for the abundance and diversity of life on Earth. There is a lot of evidence supporting the theory of evolution by natural selection. Let's look at some of that evidence. The fossil record is one important type of evidence. **Fossils** may be mineralized remains of an organism, a casting (or impression) of an organism that was once alive, or remains left behind by a living organism, such as a burrow, footprints, or waste. These fossils may be held in different substances like tar, ice, rock, volcanic ash, and amber. As the years and centuries pass, fossils become compressed under compounded and new accumulations of soil and rock. If those layers (or **strata**) remain undisturbed, fossilized evidence of life can be explored and dated based on their depth in the rock layers and other materials found in the same stratum. **Stratification** has been used to help establish the age of some of the oldest life forms on Earth. The oldest fossils on record are those of bacteria that lived 3.7 to 3.8 billion years ago. Fossilized evidence exists of 13-foot-long molluscs, called nautiloids, and 28-inch-long trilobites, both of which lived in oceans around the world from 500 to 252 million years ago.

Another category of physical evidence for evolution is **comparative anatomy**. Here, we look at similarities in the structures that are present in different species in order to try to figure out if, and how, they are related. **Homologous** and **vestigial structures** are comparative anatomical structures that are found in different organisms and may provide evidence of evolution. Homologous structures include features that different organisms have in common because their common ancestors also had those features. An example of homologous structures would be the five-fingered limbs of humans, cats, horses, bats, and dolphins; all are very different looking, but all have similar skeletal anatomy, implying common ancestry.

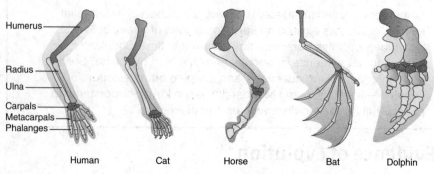

Humerus
Radius
Ulna
Carpals
Metacarpals
Phalanges

Human   Cat   Horse   Bat   Dolphin

Figure 16–5. Homologous Structures

Vestigial structures have no current function in an organism, but come from an ancestor that did have a use for that organ. As time goes by, individuals adapt and populations evolve, but the vestigial structures still remain. Some examples of vestigial structures include pelvic bones in whales, leg bones in snakes, flightless birds with wings, and cave-dwelling reptiles and fish that have eye structures even though they live in complete darkness. Vestigial structures in humans include the tailbone, wisdom teeth, and goosebumps.

Comparative embryology is the study and comparison of the fetal anatomy and development of different species. As you can see in Figure 16-6, embryos of most species look very different from the adults of their species. Although embryo sizes and developmental time frames vary, evolutionary relationships indicating common ancestry are often supported when the appearances of fetuses of different species are very similar.

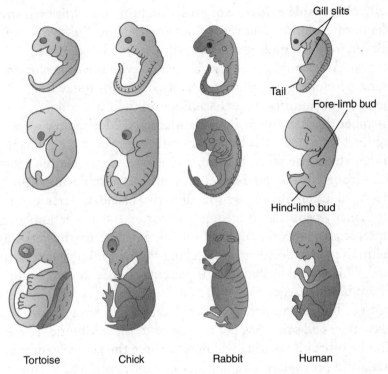

Gill slits

Tail

Fore-limb bud

Hind-limb bud

Tortoise   Chick   Rabbit   Human

Figure 16–6. Comparative Embryology

Examining physical evidence like muscle tissue, skin, nerve tissue, blood, and sensory tissue allows scientists to understand and compare **biochemical relationships** between species. When examining the cells, biochemistry, and organelles of different species, scientists find that these different tissues and molecules will often support common ancient ancestry. This is because different types of cells, organelles, and molecules may have a similar appearance, and perform the same functions. Examples of biochemical evidence of evolution include mechanisms used by cells to obtain energy, metabolize nutrients, dispose of wastes, build proteins, and maintain nucleic acids. Today, **molecular genetics** is at the core of identifying evolutionary relationships. Transcription of DNA into RNA and then RNA into specific proteins is a process common to all living things and is considered to be very strong evidence for the evolutionary relationships between all living things on Earth.

Finally, **observable evidence** of evolution provides additional strong evidence of evolution. Many organisms with short life spans and quick generational cycling can be easily observed undergoing evolution. An example of easily observable evidence of evolution is bacteria developing antibiotic resistance. Humans today have frequent opportunities to see bacterial evolution in action. Remember the last time you took antibiotics? Maybe they were prescribed for a sore throat or a skin infection. You were probably told they should be taken for five, seven, or 10 days. Antibiotics work to kill off specific bacterial infections, but they also destroy healthy bacteria, including many essential digestive bacteria. After a couple of days, many people start feeling better from their infection but get an upset stomach because of all of the missing digestive bacteria. So, sometimes people decide to stop taking their antibiotics before they should. This is a really bad idea. If patients choose to stop taking their antibiotics early, most of the infectious bacteria will already be destroyed. The remaining small percentage of bacteria are still there because they had some level of resistance to the antibiotic. These antibiotic resistant bacteria will now become the primary infecting bacteria and previously used antibiotics won't work well against them. The bacterial population has evolved.

PAINLESS FACT

Sometimes people say that they "don't believe in" evolution or that natural selection is "just a theory." These comments reflect a misunderstanding of the science that they are discussing. Let's dissect these comments so that we can effectively respond to them. First, science is not about "believing in" anything. Using science to understand the world means that we ask thoughtful questions and then search for answers through research, experimentation, and data collection. Second, the evidence for bacterial evolution in the development of antibiotic resistance is observable, documented, and supported. Third, scientific theories are not like historical or common theories. They represent the ultimate result of thoroughly tested hypotheses and experiments that are repeatedly supported over time and by many scientists. A scientific theory is so well supported that it is accepted as a foundational principle in science.

BRAIN TICKLERS   Set # 3

Match these descriptions to the correct terms.

1. Evidence of evolution involving the mineralized remains of an organism, a casting, or the remains left behind by a previously living organism

2. Evidence of evolution that we can see as it occurs

3. Evidence of evolution comparing fetal development of different species

4. Evidence of evolution involving comparisons of molecules found in the remains of previously living organisms

A. observable

B. biochemical

C. fossil

D. embryological

(Answers are on page 274.)

Human Evolution

The history of the human race began with the evolution of all life on Earth. We are able to follow the timeline of **divergent evolution** that reflects the evolution of humankind by observing the fossil record. Around 220 million years ago (mya), the earliest mammals evolved. They were small, burrowing, warm-blooded, insect eaters; they produced milk for their young; and they had brains capable of increasingly complex thinking. These mammals adapted and thrived for millions of years. Around 65 mya, a group of divergent mammals evolved into primates, with larger brains, flattened faces, and intricately functional hands and feet. Primates continued to adapt to their environments and evolve. In Africa, around 40 mya, some of these primates evolved into early monkeys, tarsiers, and apes. Some early primates made it to the New World (what we now call South America) around 30 mya by floating across the Atlantic Ocean on huge tangled mats of vegetation. They became the New World monkeys, like capuchins and woolly monkeys. Those primates that remained in Africa eventually evolved into the Old World monkeys (including baboons, mandrills, and macaques) and the earliest apes. Between 15 and seven mya, the great apes, family *Hominidae*, emerged; this family includes gorillas, chimpanzees, orangutans, and humans.

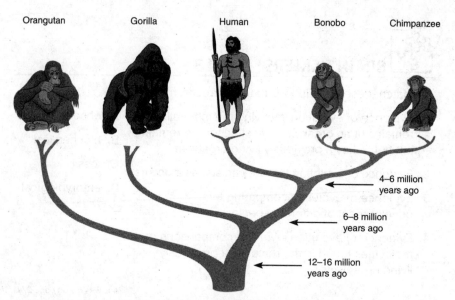

Figure 16–7. Divergent Primate Evolution

The earliest upright (**bipedal**) ancestors of the human race appeared between 3.7 and 2.5 mya and belonged to the genus *Australopithecus.* These australopithecines thrived in the savannas of Africa, becoming fully bipedal, losing their body hair, and enhancing their brain capacity. This group thrived, continued to adapt to its environment, and ultimately evolved into the earliest members of the genus *Homo.* Around 2.5 mya, selective environmental pressures were steering the ancestors of today's gorillas and chimpanzees via divergent evolution from the evolutionary direction of early human ancestors. The earliest fossil evidence of tools also appeared around 2.5 mya and was associated with an early human ancestor called *Homo habilus.*

A number of different species of ancient humans have been discovered as fossil evidence, but the line of descent does follow some significant pre-human hominids. Around 1.6 mya, in Africa, *Homo erectus* were early human ancestors who walked upright, used fire as a tool, and had flattened facial features similar to humans today. The species *Homo erectus* became extinct around 70,000 years ago. Neanderthals, or *Homo neanderthalensis,* existed from 300,000 to 27,000 years ago. The Neanderthals coexisted with modern man, *Homo sapiens.*

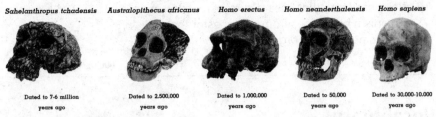

| *Sahelanthropus tchadensis* | *Australopithecus africanus* | *Homo erectus* | *Homo neanderthalensis* | *Homo sapiens* |

Dated to 7-6 million years ago · Dated to 2,500,000 years ago · Dated to 1,000,000 years ago · Dated to 50,000 years ago · Dated to 30,000-10,000 years ago

Figure 16–8. Skulls as Evidence of Human Evolution

Fossil records indicate that the earliest *Homo sapiens* lived in the region of Africa now known as Ethiopia about 195,000 years ago. Strong social connections in tribes of *Homo sapiens* are evidenced by fossil evidence of death rituals beginning around 160,000 years ago. Genes related to speech have been found in the fossilized remains of *Homo sapiens* dating back to around 70,000 years ago. The ability to communicate enhanced the ability of *Homo sapiens* to develop and expand their territory as well as their social and cultural interactions. *Homo sapiens* migrated into Asia 50,000 years ago and into Australia and Europe 40,000 years ago. *Homo sapiens* became the only members of the genus *Homo* living on Earth when the species *Homo floresiensis* became extinct 12,000 years ago. By the year 10,000 B.C.E. (before the Common Era), *Homo sapiens* were developing civilizations, domesticating animals and crops, and farming.

SUPER BRAIN TICKLERS

Match these definitions to the correct terms.

1. Human ancestor who walked upright, used fire as a tool, and had flattened facial features

2. Modern man

3. Feature that different organisms have in common because their common ancestors also had that feature

4. Human ancestor associated with the earliest fossil evidence of tools

5. Soil and rock levels in undisturbed earth

A. *Homo habilus*

B. stratification

C. *Homo sapiens*

D. vestigial structure

E. *Homo erectus*

F. homologous structure

6. Body structure that has no current function in an organism but comes from an ancestor that did have a use for that organ

(Answers are on page 274.)

Vocabulary

Adaptive radiation: The diversity of variations within a species based upon living in different environments.

Biochemical relationships: Relationships between species understood by comparing molecules and mechanisms used by cells to obtain energy, metabolize nutrients, dispose of wastes, build proteins, and maintain nucleic acids.

Bipedal: Describes organisms that can walk upright and on two feet.

Cladogram: A helpful tree-shaped diagram that shows the evolutionary relationships between living things based on both physical traits and DNA.

Comparative anatomy: Study of the similarities in the anatomical structures that are present in different species in order to try to figure out if, and how, the organisms are related.

Comparative embryology: The study of how the fetal anatomy and development of different species compare to one another.

Divergent evolution: The process of two or more related species becoming less alike as time goes by.

Evolution: The changes that happen in a population over time.

Extinct: A species that is no longer living.

Fossil: The mineralized remains of an organism, a casting (or impression) of an organism that was once alive, or the remains left behind by a living organism such as a burrow, footprints, or waste.

Homologous structures: Features that different organisms have in common because their common ancestors also had those features.

Molecular genetics: At the core of identifying evolutionary relationships, transcription of DNA into RNA and then RNA into specific proteins is a process common to all living things and is considered to be very strong evidence for the evolutionary relationships between all living things on Earth.

Observable evidence: Many populations of organisms with short life spans and quick generational cycling can be easily observed undergoing evolution. These organisms provide strong evidence of evolution.

Primates: The earliest recognizable mammalian ancestors of the human race; they are still around today. Lemurs, lorises, tarsiers, New World monkeys, Old World monkeys, orangutans, gorillas, chimpanzees, bonobos, and humans are the primates alive in the world today.

Primordial soup: The mixture of elements that existed in the oceans of ancient Earth; they were jolted with electricity, mixed violently, and heated. These elements eventually formed unique molecules and bonds that were responsible for the development of life on Earth.

Strata: Undisturbed layers of earth that may contain evidence of past lives and activities.

Stratification: In undisturbed earth, soil and rock levels that can be explored. Any embedded fossils can be dated based on their depth in the rock layers and other materials found in the same stratum.

Theory of natural selection: Darwin's explanation and recognition of the mechanism by which species evolve.

Vestigial structures: Body structures that have no current function in an organism but come from an ancestor that did have a use for those structures.

Brain Ticklers—The Answers

Set # 1, page 259

1. extinct
2. evolution
3. cladogram

Set # 2, pages 263–264

1. C
2. D
3. B
4. A

Set # 3, page 269

1. C
2. A
3. D
4. B

Super Brain Ticklers

1. E
2. C
3. F
4. A
5. B
6. D

Chapter 17

Human Impacts on Earth

Humans are a unique group of animals. We think. We use tools. We can communicate verbally as well as in writing. We have a huge vocabulary. We document our history, our accomplishments, and our failures. We use words that distinctly identify our ideas and our emotions. Humans have designed planes, cars, ships, and submarines to travel around the world, deep into the oceans, and through the skies. Humans have traveled to and walked on the moon. Humans are the one population on Earth that has the capacity to change the Earth. This is a tremendous opportunity and it's also a huge responsibility! What have we done with this amazing ability? Have we consistently worked to make the world a better place? That would make sense, right?

Unfortunately, in our pursuit of everything we can accomplish and acquire, we humans have done a lot of damage here on Earth. We've sprawled into previously unlivable environments, waged wars, misused land, destroyed ecosystems, depleted resources, clear-cut forests, and polluted the land, water, and air. It seems like a good time to consider our role as protectors of the Earth and our responsibility to manage the Earth with care.

The Anthropocene Era

Humans have changed the Earth far more than we could have imagined 100, or even 50, years ago. The human population of Earth is on the verge of 8 billion people and is continuing to grow. Forests have been leveled and grasslands have been covered with communities. Water sources are being depleted and huge quantities of carbon dioxide have been pumped into the Earth's atmosphere. Polar ice caps are melting, raising sea levels and interfering with the ocean's biochemistry. A "garbage patch" about three times the size of France is floating in the Pacific Ocean between Hawaii and California. Wild and unusual weather patterns are impacting countries around the world because of human-driven climate change. We have destroyed habitats and have moved species from their natural habitats to new environments, often disrupting normal species interactions in their new homes. Human activities have triggered the extinctions of millions of plant and animal species and are continuing on this path today. Have we, as members of the human race, reached our carrying capacity on the Earth? If not, what is the carrying capacity likely to be and what will happen when we get there? Are scientists correct in wanting to name our current geological time period the "age of humans," also known as the **Anthropocene**?

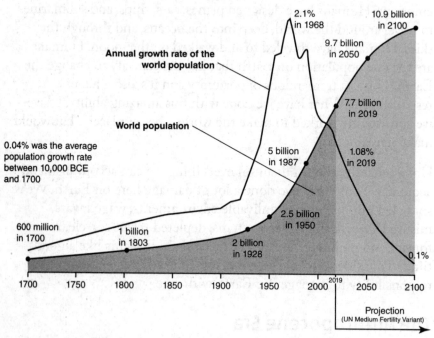

Figure 17–1. World Population Growth 1700–2100

Conservation Biology Impacts

Scientists who study the natural world with an emphasis on protecting the world's biodiversity are working in a field called

conservation biology. Conservation scientists work to protect species and ecosystems. The field of conservation biology merges the fields of ecology, marine biology, genetics, statistics, natural resource management, population biology, and economics to develop resource management plans and policy. Conservation biologists identify species that are at risk and habitats that are being damaged and destroyed. They explore all of the criteria that are likely to impact biodiversity. Some threats to biodiversity, like **geographic isolation** and natural disasters, occur without human interference. Some threats are very clearly the result of human plundering of the Earth's resources—for example, humans interfering with pristine ecosystems, and **poaching** and removing exotic pets from their natural environments. Other threats to biodiversity are almost entirely man-made and include climate change, habitat destruction, pollution, deforestation, urbanization, and population sprawl. In order to understand conservation biology and identify how we can best serve and protect our planet, we need to consider all of the problems we've created.

 BRAIN TICKLERS   Set # 1

Select the correct term to complete the sentence.

1. Many elephants in Africa are killed so that hunters can take their tusks in a crime called (<u>encroaching</u> or <u>poaching</u>).

2. Scientists who study the Earth's resources and how they relate to protecting biodiversity are called (<u>conservation</u> or <u>ecological</u>) biologists.

3. Some scientists are proposing that we call the current time period on Earth the (<u>Humanocene</u> or <u>Anthropocene</u>) Era.

4. (<u>Habitat</u> or <u>Geographic</u>) isolation occurs when members of a population become physically separated, often because their original habitat is divided.

(Answers are on page 287.)

The introduction of pollutants, or harmful contaminants, into the environment can occur naturally or can be man-made and is called **pollution**. Often, pollution is identified based on the part of the environment that it impacts: water, soil, or air. Pollution exists in many different forms, including the overuse and mismanagement of fertilizers and agricultural products, toxic chemical releases, and mismanaged emissions and waste from factories, power plants, homes, vehicles, and poorly managed garbage dumps.

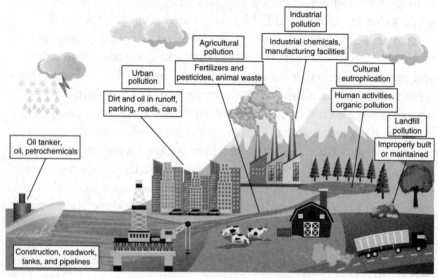

Figure 17–2. Sources of Pollutants

Pollutants that are released from a clearly identifiable source are called point source pollutants, while those that are dispersed from a less clear, more indirect source are called non-point source pollutants.

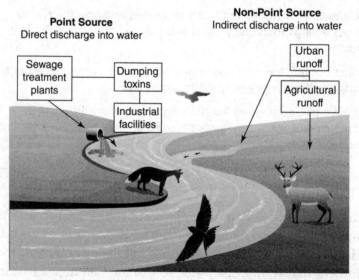

Figure 17–3. Point Source and Non–point Source Pollution

As the human population has increased, people have moved into, and disturbed, previously pristine natural environments. City boundaries have expanded farther and farther into their suburban, rural, and uninhabited neighboring areas, causing **urban sprawl**. In addition to simple overcrowding, the human population density in urban environments can generate waste disposal problems as well as high levels of noise and light pollution.

The release of gases into the atmosphere is a natural process. During the cycling of photosynthesis and cellular respiration, oxygen and carbon dioxide are continuously being released into the air. A healthy atmosphere on Earth is made up of 78% nitrogen, 21% oxygen, 0.93% argon, and 0.04% carbon dioxide along with trace amounts of neon, helium, methane, krypton, hydrogen, and water vapor. Our atmosphere is thrown out of balance when humans burn **fossil fuels** like gas or coal, which are naturally occurring fuels made of the decomposed remains of previously living organisms. We often use fossil fuels to provide power to keep factories and households running, to provide heat and cooling in our homes, and to power our trucks, cars, trains, and planes. The emissions produced by burning fossil fuels adds chemicals to the atmosphere that interfere with the normal, healthy atmospheric chemical balance. These chemicals are called **greenhouse gases** and include excess carbon dioxide, methane, nitrous oxide, ozone, and excess water vapor. Increases in greenhouse gases trigger a number of effects on Earth that are collectively called **climate change**. Increased atmospheric carbon dioxide contributes to **global warming** and **acid rain**, as well as ocean acidification and warming. All of the gases in the atmosphere interact with the infrared spectrum. Each element absorbs different wavelengths of light and causes an increase in the atmospheric insulation layer surrounding the Earth. This insulation layer is normal, however; as more greenhouse gases accumulate, it begins to insulate too much, resulting in ocean and global warming patterns. Warming of oceans contributes to melting of polar ice, which can cause ocean level rises and coastal flooding, which then jeopardizes polar and flood-prone ecosystems. The interaction of carbon dioxide with water triggers the release of hydrogen ions, resulting in acid rain and increased ocean acidity. When a water-based ecosystem is contaminated by acidity, the living things in that habitat (including fish, plankton, algae, trees, and corals) are often damaged or killed. Scientists have concluded that global warming is associated with extreme weather patterns, severe climate changes, rising sea levels, increased wildfire risk, and increasing size of deserts.

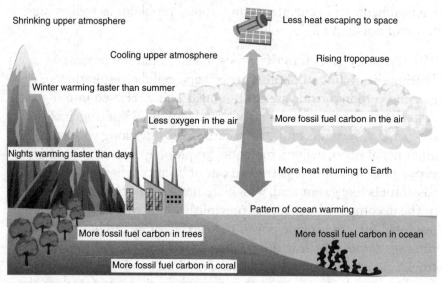

How We Know We're Causing Global Warming

Shrinking upper atmosphere

Less heat escaping to space

Cooling upper atmosphere

Rising tropopause

Winter warming faster than summer

Less oxygen in the air

More fossil fuel carbon in the air

Nights warming faster than days

More heat returning to Earth

Pattern of ocean warming

More fossil fuel carbon in trees

More fossil fuel carbon in ocean

More fossil fuel carbon in coral

Figure 17–4. Indicators of Global Warming

Another area of concern for conservation biologists involves poor resource management and wasted resources. Habitat destruction, **deforestation**, depletion of fossil fuels, giant ocean "garbage patches," and **desertification** are all examples of inadequate resource management and wasteful use of resources. Deforestation results from the clear-cutting of trees to make way for human activities like farming and urban expansion; from widespread, unmanaged logging; and from the burning of trees. Deforestation of rain forests is especially concerning because healthy rain forests remove an enormous amount of carbon dioxide from the atmosphere in order to conduct photosynthesis and, in so doing, reduce the amount of excess carbon dioxide entering the atmosphere as a greenhouse gas. The decimation of forests worldwide has been associated with soil erosion and desertification. Desertification is the conversion of once productive land into desert. Without careful conservation and management, many deforested ecosystems become wastelands, habitats are destroyed, and biodiversity is lost.

PAINLESS FACT

The government of Brazil has reported that deforestation of the Amazon rain forest increased by 22% between August 2020 and July 2021, the highest rate of deforestation in 15 years. Approximately 5,100 square miles of rain forest (an area approximately the size of Connecticut) were lost in one year. Impacted by these losses are around one million indigenous people and about three million species of plants and animals. These numbers are a significant indicator of the importance of finding ways to monitor and manage our worldwide ecological resources more effectively than we have in the past.

BRAIN TICKLERS   Set # 2

Match these descriptions to the correct terms.

1. Gas, oil, and coal, which are made of the decomposed remains of previously living organisms

2. The change from once productive land into dry and unusable land

3. The expansion of highly populated cities into neighboring areas

4. Effects on Earth such as extreme weather patterns, rising sea levels, and increased wildfire risk, which are a result of increasing greenhouse gases

A. fossil fuels

B. urban sprawl

C. climate change

D. desertification

(Answers are on page 287.)

Habitat destruction is a significant factor in biodiversity loss. Species often become endangered or extinct when their habitat is destroyed. When members of a species no longer have the water, food, or shelter that they need, they will either adapt to their new circumstances, emigrate to a new habitat, or experience a die-off. Any species that is experiencing a severe population decline and is threatened with the possibility of extinction is considered to be **endangered**. The lemur, black rhinoceros, mountain gorilla, Siberian tiger, and leatherback sea turtle are all severely endangered species. When a species no longer exists, it is called **extinct**. Examples of extinct species include the dodo, passenger pigeon, woolly mammoth, and *Tyrannosaurus rex*.

Significant conservation efforts are impacting the fossil fuel industry. Nonrenewable fuels like coal, oil, and gas have not been managed well over the last 150 years and, because they can't be regenerated, at the rate we are using them, they are in danger of being completely depleted by the end of this century. Scientific development of renewable energy sources has provided important alternatives to the decimation of our nonrenewable resources. Hybrid and battery-operated cars are gaining in number and popularity worldwide, with some governments and automakers committing to manufacturing significantly fewer gas-powered cars over the next 5–20 years. Renewable alternative energy sources include solar, wind, geothermal, hydroelectric, tidal, wood, nuclear, and waste-to-energy.

Some of the environmental issues that conservationists work with involve individual decisions, cultural norms, political manipulation, and legal maneuvering. Individuals can play a role in strengthening the health of our planet by reducing their use of disposable materials and by opting to **recycle** and properly get rid of waste. People can also choose to help in conservation efforts by choosing personal pets and landscaping plants carefully. Nontraditional and exotic pets and plants may be interesting and fun to have, but if they get loose and spread, they can decimate local or **native species**. **Invasive species** include any species that did not originate in the habitat where it has been deposited. They are often able to overpopulate and, without predators or inhibiting factors, can negatively alter their new environment. Overpopulation of boa constrictors (large snakes) in Florida and kudzu (dense leafy plants) throughout many American southeastern states are both examples of invasive species having a negative impact on their adopted ecosystems.

Human Intellect Impacts

Humanity has had quite an impact on the planet we inhabit. As individuals, we constantly make choices that impact the Earth. Over the centuries, and especially over the last 200 years, humankind has often taken the resources on Earth for granted. We have often used and abused the environment. Today, many people have recognized this and have begun to reduce their negative footprint, or impact, on the Earth. These efforts encourage **sustainability**. Sustainability means living in a way that minimizes human impacts on the environment and on the Earth. Living sustainably can involve a personal reprioritization of our lifestyle—perhaps eating and using fewer animal products, planting native species in our yards and neighborhoods, or using cars less and trying to walk, bike, or use public transportation more. It may involve sharing information about sustainability and working toward cooperative efforts with your family, neighborhood, and community. A commitment to sustainable living can involve the use of green technologies and renewable energy in a home, use of sustainable transportation, organic agriculture and gardening, and sustainable architecture.

PAINLESS FACT

While there is a scientific field and career path called conservation biology, it doesn't mean that we can't all act as conservationists as we live our daily lives. Is there some aspect of helping the Earth to stay "healthy" that you feel passionate about? Have any of the conservation issues in this chapter sparked your interest? If so, the world needs you. Learn all you can about that topic, develop your skills and network, commit to spending time and volunteer, and look at what other people are doing and sharing to solve that problem. Collaborate. If each of us contributes what we are capable of giving, we can make a difference.

We humans are a powerful species. We have developed ways to travel throughout the Earth and into space. We have mapped our DNA as well as the distant planets. We have harnessed power from atoms, from molecules, from wind, from lakes, and from the sun. We have controlled and domesticated animals much larger and more powerful than ourselves. We have created and destroyed great civilizations. We have developed powerful technology that can store information in the "cloud," generate artificial intelligence, and allow instantaneous worldwide communication.

All of these accomplishments have been possible because of our most powerful tool—our brains. We each have a choice and a responsibility to explore and try to understand ourselves, each other, and the world we share. We each have the potential and the responsibility to impact the world in a constructive and positive way.

SUPER BRAIN TICKLERS

Match these definitions to the correct terms.

1. Living in a way that minimizes human impacts on the environment

2. Convert waste into a reusable material

3. A species that is experiencing a severe population decline and is threatened with the possibility of dying off

4. A species that no longer exists

5. An environmental chemical reaction that happens because hydrogen ions are released when carbon dioxide reacts with water

6. Species that did not originate in the habitat where they have been deposited and which can negatively alter their new environment

A. acid rain

B. endangered

C. extinct

D. invasive

E. sustainable

F. recycle

(Answers are on page 287.)

Vocabulary

Acid rain: An increase in water's acidity as a result of hydrogen ions being released when carbon dioxide reacts with water. This is common today because CO_2 is a greenhouse gas.

Anthropocene: A suggested name for the current geological time period: the "age of humans."

Climate change: Effects such as extreme weather patterns, rising sea levels, increased wildfire risks, and increasing size of deserts, triggered by increases in greenhouse gases.

Conservation biology: The study of the Earth's resources and how they relate to protecting biodiversity.

Deforestation: The loss of forested ecosystems resulting from the clear-cutting of trees; widespread, unmanaged logging; and the burning of trees.

Desertification: The conversion of once productive land into desert.

Endangered species: Any species that is experiencing a severe population decline and is threatened with the possibility of extinction.

Extinction: When a species dies off; no longer exists.

Fossil fuels: Gas, oil, and coal; naturally occurring fuels made of the decomposed remains of previously living organisms.

Geographic isolation: Occurs when members of a population become physically separated, often because their original habitat is divided; this can happen as a result of the formation of new water or land barriers.

Global warming: As greenhouse gases accumulate, they begin to insulate Earth's atmosphere too much, resulting in ocean and global warming patterns.

Greenhouse gases: Emissions produced by burning fossil fuels, adding excess chemicals (including carbon dioxide, methane, nitrous oxide, ozone, and water vapor) to the atmosphere. These chemicals interfere with the normal atmospheric chemical balance.

Invasive species: Species that did not originate in the habitat where they have been deposited. They are often able to overpopulate and, without predators or inhibiting factors, can negatively alter their new environment.

Native species: Organisms that live in the habitat where their ancestors have lived. They have a niche in that ecosystem and tend to have a healthy and balanced population.

Poaching: The illegal hunting or capturing of wild animals.

Pollution: The introduction of harmful contaminants into the environment.

Recycle: Convert waste into a reusable material.

Sustainability: Living in a way that minimizes human impacts on the environment and on Earth.

Urban sprawl: Expansion of highly populated cities into their suburban, rural, and uninhabited neighboring areas.

Brain Ticklers—The Answers

Set # 1, page 277

1. poaching

2. conservation

3. Anthropocene

4. Geographic

Set # 2, page 281

1. A

2. D

3. B

4. C

Super Brain Ticklers

1. E

2. F

3. B

4. C

5. A

6. D

Index

Note: Page numbers followed by "f" represent figures.

abiotic factors, 213, 222
abyssal zone, 228
accessory organs, 122–123
acid rain, 279, 284
acquired immunity, 121
activation energy, 45, 48
active transport, 61–62,
    62f, 64
adaptation, 4
adaptive radiation, 263, 272
adenosine diphosphate
    (ADP), 24, 29, 46,
    81, 81f
adenosine triphosphate
    (ATP), 24, 29, 46–47,
    77, 81, 81f
    defined, 46, 48
    importance of, 47
aerobic respiration, 77–78,
    78f, 82
alcoholic fermentation,
    79, 82
alimentary canal, 108
alleles, 149, 150, 150f
    heterozygous, 156, 164
    homozygous, 156, 165
    multiple, 163, 165
Alternaria spp., 21
Alzheimer's disease, 193, 197
amino acids, 37, 42, 48
    abbreviations for, 178f
amniocentesis, 195, 197
amoebas, 20
anabolism, 37, 48
anaerobic respiration, 77,
    79–80, 80f, 82
anaphase, 132, 139
anatomy, 87
    macroanatomy, 90
    microanatomy, 90
animals, cells of, 21–22
anterior side, 88, 124
Anthropocene era, 275–276,
    276f, 284
anticodon, 175, 179
apex predators, 219, 222

apoptosis, 27, 29, 136, 139
appendages, 100
appendicular skeleton, 93
Archaea, 18, 19
arterial system, 104f
arteries, 103
articulations, 93, 124
-ase (suffix), 45, 172
asexual reproduction, 3, 129,
    139
asthma, 107
atherosclerosis, 105
athletes, 69
at-home DNA test, 202–203
atmosphere, 231
atoms, 33, 48
Australopithecus, 270
autoimmune diseases, 193,
    197
autonomic nervous
    system, 114
autotrophs, 21, 24, 29, 70,
    82, 214, 214f, 222
Avery, Oswald, 167
axial skeleton, 93

bacillum, 19
bacteria, 18, 19–20, 147, 219
bar graph, 247
benign prostatic
    hypertrophy, 123
benthic zone, 228
bias, 6, 11
binary fission, 19, 29
biochemical relationships,
    267, 272
biodiversity, 276, 277
biological chemistry, 33–50
    atoms, 33
    chemical bonds, 36–37
    molecules, 35, 45–46, 46f
    water, 37–38
biological macromolecules,
    39–44
    carbohydrates, 39, 40, 40f
    lipids, 39, 41, 41f

nucleic acids, 39, 42–43, 44f
    proteins, 39, 41–42, 42f
biological taxonomy, 18
biology, 2–3
    methods to explore, 5–11
    recognition of
        characteristics of life, 3–4
    as science of life, 2–3
biomes, 228–231, 241
    temperate forest, 229–230
    terrestrial, 229
biosphere, 231, 232f, 241
biotic factors, 213, 214, 222
    autotrophs, 214, 214f
biotic potential, 246, 253
bipedal, 270, 272
bleeding, 191–192
blood, 101, 102
    cells, 91
    vessels, 99
bones, 91. *See also* skeletal
        system
    long, 93
    microscopic anatomy of,
        92, 92f
    short, 93
Bowman's capsule, 111
Brazil, 281
breathing. *See* respiratory
        system

calcium ions, 61
cancer, 27, 193, 197
capillaries, 103
carbohydrates, 39, 40, 40f,
    48, 55, 70
carbon cycle, 238–239, 239f
cardiac muscle, 95, 96f
cardiovascular system,
    101–105
    disorders, 104–105
    functions of, 101
    macroscopic anatomy,
        102–103
    microscopic anatomy,
        101–102

carnivores, 218, 222
carrying capacity, 246, 253
cartilage, 93
catabolism, 37, 48
catalyze, 45
cavities, human body,
    89, 89f
cell cycle, 130–131,
    130f, 139
  interphase, 130
  meiosis, 130, 131,
    133–135, 134f
  mitosis, 130, 131–133,
    132f
cell membrane, 22–23, 29
  improper working, 63
  structure, 53–55, 55f
  transport controlled by, 56
cells, 4
  animals, 21–22
  archaeal, 19
  bacterial, 19–20
  blood, 91
  defined, 15–16
  diploid, 133
  diversity of, 19–22
  eukaryotes, 18–19, 53
  fungi, 21
  glial, 114, 114f
  haploid, 134
  human, types of, 16
  improper working of, 27
  nerve, 113–114
  parts of, 22–25, 26f
  plant, 21
  prokaryotic, 18, 19
  reproduction. See cellular
    reproduction
  sizes and shapes, 15, 16f
  as structural units, 15
cellular immortality, 136, 139
cellular reproduction,
    129–140
  asexual reproduction, 129
  cell cycle. See cell cycle
  sexual reproduction,
    129–130
  telomeres, 136–138, 137f
cellular respiration, 70,
    76–80, 82

aerobic respiration,
    77–78, 78f
anaerobic respiration, 77,
    79–80, 80f
cellular transport, 53
  active transport, 61–62,
    62f
  cell membrane and, 56
  passive transport, 57–60
cell wall, 25, 29
central nervous system
    (CNS), 114
centromere, 131, 139
Chargaff, Erwin, 168
chart, 247
Chase, Martha, 168
chemical bonds, 36–37
  defined, 36
  importance of, 37
  types of, 36–37
chemical energy, 73
chemistry, 2
  biological. See biological
    chemistry
chemosynthesis, 75, 76f
chlorophyll, 72, 82
chloroplasts, 24, 29, 72, 83
cholecystitis, 110
CHONPS mnemonic, 34
chorionic villus sampling
    (CVS), 195, 197
Chromista, 20–21
chromosomal abnormalities,
    194–195
chromosomes, 147–148,
    150, 170, 179
  homologous, 149
cilia, 21, 25, 29
cladogram, of life on Earth,
    258f, 259, 272
clarity, data, 9
climate change, 227,
    279, 284
climax community,
    233, 241
cloning, 208–209, 210
codominance, 162–163,
    162f, 164
codons, 175, 177f, 178, 179
commensalism, 221, 222

communication, data, 7–9
communities, 227, 241
  climax, 233
  and succession, 232–234
comparative anatomy,
    266, 272
comparative embryology,
    266, 267f, 272
competition, 249
complementary DNA
    strand, 170, 179
compound light microscope,
    9–10, 10f, 12
concentration gradient,
    61, 64
  molecules travel against,
    61
conservation biology,
    276–282, 284
consumers, 70, 83, 214,
    214f, 222
control variables, 7–8, 12
COPD, 107
coral reefs, 228–229
covalent bonds, 36,
    36f, 48
COVID-19, 204, 205
C3 photosynthesis, 74–75
C4 photosynthesis, 74–75
Crick, Francis, 168–169
CRISPR (gene editing
    tool), 207, 208f, 210
Crohn's disease, 110
crossing over, 135, 139
Cushing's syndrome, 119
cyanobacteria, 72, 83
cystic fibrosis (CF), 27, 63,
    188–189, 197
cystic fibrosis transmembrane
    conductance regulator
    (CFTR) gene, 188
cytokinesis, 132, 139
cytoplasm, 23, 30
cytosines, 174
cytoskeleton, 25, 30

Darwin, Charles, 262–263
  theory of natural
    selection, 264–265,
    264f

data
  communication, 7–9
  defined, 6, 12
  qualitative, 6, 13
  quantitative, 6, 13
decomposers, 218, 219, 222
deforestation, 280, 281, 285
dehydration synthesis, 38,
  38f, 48
demographic transition,
  252, 252f, 253
demography, 251–252,
  252f, 253
density-dependent limiting
  factors, 247–249, 254
density-independent
  limiting factors, 247,
  249–250, 254
dependent variable, 8, 12
dermis, 99, 99f
desertification, 280, 285
detritivores, 218, 222
development, as living
  things characteristic, 3
diabetes, 189, 197
diabetes mellitus, 118
diarthrotic joints, 93
diffusion, 57, 58f, 64
  facilitated, 60, 60f
digestive system, 108–110,
  110f
  diseases/disorders, 110
  functions, 108
  macroscopic anatomy,
  108–109
  microscopic anatomy, 108
dihybrid, 157, 164
diploid, chromosomes, 147,
  151
diploid cells, 133, 139
directional terms, 88, 88f
distal, 88, 124
divergent evolution, 269,
  270f, 272
DNA (deoxyribonucleic acid),
  42–43, 44f, 48, 137, 144,
  146, 167–168, 257–258
  at-home test, 202–203
  chips/microarrays, 202,
  210

molecule, structure of,
  170, 170f
DNA replication, 172–178,
  179
domains, 17, 18–19
dominant trait, 156, 164
dorsal cavity, 89, 124
Down syndrome, 194
Duchenne's muscular
  dystrophy (DMD), 97,
  192, 198
dynamic equilibrium, 57, 64

Earth
  biomes, 228–231
  cladogram of life on,
  258f, 259
  communities and
  succession, 232–234
  ecological levels of
  organization, 225–228,
  226f
  evidence of life on,
  259–262, 260f
  human evolution on,
  269–271, 270f, 271f
  human impacts on,
  275–286
  nutrient cycle, 237–239
ecology
  biotic factors. See biotic
  factors
  defined, 213, 222
  feeding relationships, 216
  food webs, 214, 217–219,
  217f
  habitat, 215, 215f
  population biology and.
  See population
  relationships between
  species, 213–214
  species interactions,
  219–220
  symbiotic relationships,
  221
ecosystem, 215, 222, 228,
  241
  biomes, 228–231
  nutrient cycling in,
  237–239, 240f

water-based, 279–280
eczema, 100
electron microscopes,
  10–11, 12
electron transport chain,
  73, 83
elements
  periodic table of, 33–34,
  34f
emigration, 246, 254
endangered species, 281,
  285
endocrine glands, 117, 118
endocrine system, 116–119,
  118f
  diseases, 118–119
  functions, 116
  macroscopic anatomy,
  117
  microscopic anatomy, 117
endocytosis, 60f, 62, 64
endometriosis, 124
endoplasmic reticulum,
  23, 30
energy
  activation, 45, 48
  chemical, 73
  exchange within
  environment, 70–72, 71f
  flow, ecological
  relationships and,
  235–236
  metabolism and, 69–70
  pyramid of, 236, 236f
energy-storing macromole-
  cules, 80–81
environment
  energy exchange within,
  70–72, 71f
enzymes
  defined, 45, 48
  importance of, 46
  structure of, 46
epidermis, 99, 99f
epidermolysis bullosa (EB),
  100
epilepsy, 116
epistasis, 163, 164
equilibrium, 59, 64
essential amino acids, 42

estuaries, 229
Eukarya, 18–19
eukaryotes, 18–19, 20–22, 30, 53, 147
evolution, 258, 272
   evidence of, 265–268
   human, 269–271, 270f, 271f
exocrine gland, 117
exocytosis, 60f, 62, 62f, 64
expiration, 105, 125
extinct, 257, 272, 281–282, 285

facilitated diffusion, 60, 60f
fallopian tubes, 123
fauna, 229, 241
Felis catus, 17
female reproductive system, 123, 123f
fertilization, 129, 139, 147
fibrodysplasia ossificans progressiva (FOP), 97–98
filial, 145
flagella, 21, 25, 30
flora, 229, 241
fluid mosaic model, 55, 65
food webs, 214, 217–219, 217f, 223
fossil fuels, 279, 285
fossils, 265, 271, 272
Franklin, Rosalind, 168
frontal plane, 88, 125
fungi, 219
   cells, 21

gamete, 129–130, 139
gametogenesis, 136f
genes, 148–149, 148f, 151, 156, 170, 180
genetically modified organism (GMO), 206–207, 210
genetic counseling, 196, 198
genetic disorders, 183–199
   basics of, 183–184
   chromosomal abnormalities, 194–195
   cystic fibrosis, 188–189
   Duchenne's muscular

   dystrophy, 192
   hemochromatosis, 189
   hemophilia, 191–192
   Huntington's disease (HD), 185–186
   Marfan syndrome, 186–187
   polygenic and multifactorial disorders, 192–193
   sex-linked disorders, 191
   sickle cell disease, 189
   Tay-Sachs, 190
genetic engineering, 205–207, 210
genetics, 143–144, 151
   alleles, 149, 150f
   chromosomes, 147–148
   defined, 143
   disorders. *See* genetic disorders
   DNA, 146
   genes, 148–149, 148f
   history of study of, 167–169
   Mendelian. *See* Mendelian genetics
   molecular. *See* molecular genetics
   pedigree, 144–145, 145f
   technology. *See* genetic technology
   traits, 144–145
genetic screening, 195–196
genetic technology, 201–210
   at-home DNA test, 202–203
   cloning, 208–209
   DNA microarrays, 202
   genetic engineering, 205–207
   mRNA vaccines, 204–205
   polymerase chain reaction, 201–202
genome, 149, 151
genotype, 156, 157, 164
geographic isolation, 277, 285

glands
   endocrine, 117, 118
   exocrine, 117
glial cells, 114, 114f
global warming, 279, 280f, 285
glomerulonephritis, 113
glomerulus, 111
glycolipids, 55
glycolysis, 72, 77, 83
glycoproteins, 55
Golden Rice, 206
Golgi apparatus, 24, 30
grana, 73, 83
graphs
   axes, 8
   bar, 247
   defined, 7, 12
   line, 7, 7f, 8, 247
grasslands, 229
greenhouse gases, 279, 285
Griffith, Frederick, 167
gross anatomy, 90, 100, 125
growth, as living things characteristic, 3
guanines, 174

habitats, 215, 215f, 223, 227–228, 281
hair, 100
haploid, chromosomes, 147, 147f, 151
haploid cells, 134, 139
haversian systems, 92
heart. *See* cardiovascular system
heart disease, 192–193
hematopoiesis, 93, 125
hemochromatosis, 189, 198
hemophilia, 104, 191–192, 198
hepatitis, 110
herbivores, 218, 223, 248–249
herbivory, 248, 254
hereditary heart disease, 198
hereditary hemochromatosis (HH), 189
heredity, 143, 151
Hershey, Alfred, 168

heterotrophic, 20, 30
heterotrophs, 70, 83, 214, 214f, 223
heterozygous alleles, 156, 164
HEXA gene, 173, 190
homeostasis, 3–4, 12
Homo, 261, 270
Homo erectus, 270
Homo habilus, 270
homologous chromosomes, 149, 151
homologous structure, 266, 266f, 272
Homo neanderthalensis, 270
Homo sapiens, 226, 262, 270, 271
homozygous alleles, 156, 165
hormones, 117, 173
human body, 87
    anatomy, 87
    cardiovascular system, 101–105
    cavities, 89, 89f
    digestive system, 108–110
    directional terms, 88, 88f
    endocrine system, 116–119, 118f
    immune system, 120–121, 121f
    integumentary system, 98–100
    lymphatic system, 119–120, 120f
    macroanatomy, 90
    membranes, 89, 90f
    microanatomy, 90
    muscular system, 95–98
    nervous system, 113–116, 115f
    physiology, 87
    planes, 88, 88f
    reproductive system, 122–124, 123f
    respiratory system, 105–107
    skeletal system. *See* skeletal system
    urinary system, 110–113

human evolution, 269–271, 270f, 271f
Human Genome Project, 149
human impacts on Earth, 275–286
    Anthropocene era, 275–276, 276f
    conservation biology impacts, 276–282
    human intellect impacts, 282–284
human population growth, 250–252, 250f
Huntington's disease (HD), 185–186, 198
hydrogen bonds, 37
hydrogen peroxide (H_2O_2), 25
hydrolysis, 38, 38f, 49
hydrophilic, 23, 54, 65
hydrophobic, 23, 54, 65
hydrosphere, 231
hyperthyroidism, 119
hypertonic solution, 59, 59f, 65
hypodermis, 99, 99f
hypothesis, 2, 12
hypotonic solution, 59, 59f, 60, 65

idiopathic, 94, 125
immigration, 246, 254
immortality, cellular, 136, 139
immune system, 120–121, 121f
    diseases, 122
incomplete dominance, 161–163, 162f, 165
independent variable, 8, 12
induced-fit model, 45, 49
inference, 6, 12
inferior section, 88, 125
inflammatory bowel disease (IBD), 110
information, identification of, 6
insertion, 96–97, 125
inspiration, 105
integumentary system. *See* skin/integumentary system

intercalated discs, 95, 125
intercellular transport, 53, 65
interphase, 130, 139
interspecific relationships, 220, 223
interstitial fluid, 119–120
intertidal zone, 228
intracellular transport, 53, 65
intraspecific relationships, 220
invasive species, 282, 285
invertebrates, 22
investigation
    process of, 1
    technology use for, 9–10
ion, 36, 49
ionic bonds, 36, 36f, 49
isotonic solution, 59, 59f, 65

J-curve of population growth, 246, 246f
joints, 91, 93

karyotype, 195, 196f, 198
keystone species, 219, 223
kidney failure, 112
kidneys, 111
Klinefelter syndrome, 194–195, 198
Krebs cycle, 72, 77, 83

lacteals, 108
lactic acid fermentation, 79, 83
lactose, 45
lateral, 88, 125
law, 2, 12
law of independent assortment, 157, 165
law of segregation, 157, 165
lemmings, 248
lichens, 233
life on Earth
    cladogram of, 258f, 259
    evidence of, 259–262, 260f

ligaments, 93, 125
light-dependent reactions, 71, 73–74, 83
light-independent reactions, 71, 74, 83
limiting factors, 247, 254
line graphs, 7, 8, 8f, 247
Linnaeus, Carl, 18, 19
lipids, 39, 41, 41f, 49
lithosphere, 231
living things
    characteristics of, 3–4
    importance of water for, 38–39
long bones, 93
lymphatic system, 119–120, 120f
    diseases, 122
lymphedema, 122
lysosome, 24, 30

macroanatomy, 90
mammals, heart in, 102
Marfan syndrome, 186–187, 198
marshes, 229
measurement, data, 9
medial, 88, 125
meiosis, 130, 131, 133–135, 134f, 139
membranes, human body, 89, 90f
Mendel, Gregor, 143, 153–155, 157
Mendelian genetics/ inheritance, 153–165
    complex patterns, 161–163, 162f
    Punnett squares, 158–160, 158f, 160f, 161f
    simple, 155–157
menstruation, 123
messenger RNA (mRNA), 170, 171f, 174–175, 176
    vaccines, 204–205, 210
metabolism, 4, 12, 37, 49
    and energy, 69–70
metaphase, 132, 140
metastasis, 193, 198
metric system, 9, 9f

microanatomy, 90
microscopes
    compound light, 9–10, 10f, 12
    electron, 10–11, 12
mitochondria, 24, 30
mitosis, 130, 131–133, 132f, 140
mnemonic, 17
molecular genetics, 167–180, 267, 273
    DNA replication, 172–178
    nucleic acids, 169–171
molecules, 35, 45–46, 46f, 49, 225
    travel against concentration gradient, 61
monohybrid cross, 157, 165
monomers, 40, 49
mortality rate, 246, 254
motor neuron system, 114
mules, 226
multiple alleles, 163, 165
multiple sclerosis (MS), 116
murder, 227
muscular system, 95–98
    diseases and disorders, 97–98
    functions of, 95
    macroscopic anatomy, 96–97, 97f
    microscopic anatomy, 95–96, 96f
mutualism, 221, 223
Mycobacterium tuberculosis, 20
myocardial infarction (MI), 105

Naegleria fowleri, 20
nails, 100
native species, 282, 285
natural immunity, 120–121
natural selection, 130, 140
nephron, 111, 111f
nerve cells, 113–114
nervous system, 113–116, 115f

diseases, 116
functions, 113
macroscopic anatomy, 114
microscopic anatomy, 113–114
neuromuscular junction, 96, 125
neurons, 113–114, 114f
neurotransmitter, 114
neutrons, 33
niche, 215, 215f, 223
nitrogen cycle, 239, 240f
nondisjunction, 194, 198
nonessential amino acids, 42
nuclear membrane, 23, 30
nucleic acids, 39, 42–43, 44f, 49, 169–171, 180
nucleolus, 23, 30
nucleoplasm, 23, 30
nucleotides, 169–170, 169f, 180
nucleus, 23, 30
nutrient cycle, 237–239, 241

obesity, 193
oblique plane, 88, 125
observable evidence of evolution, 268, 273
observation, 6, 12
oceans, regions of, 228–229
omnivores, 218, 223
On the Origin of Species (Darwin), 262
organelles, 22–25, 30, 53, 225
    diseases, 27
organization, 3, 13, 18, 87
    levels of, 225–228, 226f
origin, 96–97, 125
origins of replication, 172, 180
osmosis, 58, 58f, 65
osmotic solutions, 59, 59f
osteoarthritis, 94
osteoblasts, 92
osteoclasts, 92
osteocytes, 92
osteogenesis imperfecta, 95
osteons, 92

osteoporosis, 94
overcrowding, 249
oxidative phosphorylation,
    72, 77, 83
oxygen debt, 96, 125–126

parasitism, 221, 223, 249
parietal membranes, 89, 90f,
    126
Parkinson's disease, 116
passive transport, 57–60
    defined, 57, 65
    diffusion, 57, 58f
    facilitated diffusion, 60,
        60f
    osmosis, 58, 58f
Pauling, Linus, 168
pedigree, 144–145, 145f, 151
pelagic zone, 228
penis, 123
peptide bonds, 37
periodic table of elements,
    33–34, 34f
periosteum, 92
peripheral membrane
    proteins, 54–55
peripheral nervous system
    (PNS), 114
permafrost, 230, 241
peroxisomes, 25, 30
pH
    importance of, 39
    scale, 39f
    water as, 38
phagocytosis, 62, 65
pharmacogenomics, 202, 210
phenotype, 156, 157, 165
phospholipid bilayer, 53–54,
    54f
photolysis, 73, 84
photosynthesis, 70–72, 71f,
    72, 72f, 84
    C3, 74–75
    C4, 74–75
    light-dependent
        reactions, 71, 73–74
    light-independent
        reactions, 71, 74
physics, 2
physiology, 87

pinocytosis, 62, 65
pioneer species, 233, 241
pitcher plant catches, 248
plague, 227
planes, 88, 88f
plant cells, 21
plastids, 72, 84
platelets, 102
PMAT acronym, 133,
    134–135
pneumonia, 107
poaching, 277, 285
polar bodies, 135, 140
pollutants, 277–278, 278f
pollution, 277–278,
    278f, 285
polycystic kidney disease
    (PKD), 112
polygenic inheritance,
    163, 165
polymerase chain reaction
    (PCR), 201–202, 210
polymers, 40, 49
population, 227, 241
    growth, 245–252
pores, 100
posterior side, 88, 126
predation, 214, 221, 223
predator–prey population
    growth curves, 248
prickle, 227
primary succession,
    233–234, 234f, 241
primates, 258, 273
primordial soup, 259, 273
producers, 70, 84, 214, 214f,
    217–218, 223
products, 35, 49
prokaryotic cells, 18, 19, 30
prophase, 131–132, 140
proteins, 39, 41–42, 42f, 50,
    173
    in cell membranes, 54–55
    transport, 60
Protista, 20
protons, 33
proximal, 88, 126
Punnett squares, 158–160,
    158f, 160f, 161f, 185,
    185f, 188, 188f

pyramid of biomass, 235,
    235f, 241
pyramid of energy, 236,
    236f, 242
pyramid of numbers, 236,
    236f, 242

qualitative data/evidence,
    6, 13
quantitative data/evidence,
    6, 13

rain forest biome, 230
reactants, 35, 49, 50
recessive traits, 156, 165
recycle, 282, 286
red blood cells (RBCs), 101,
    189
    normal vs. sickle, 183f
red bone marrow, 93
regeneration, 129, 140
replication bubbles, 172, 180
reproduction, as living
    things characteristic, 3
reproductive system,
    122–124, 123f
    diseases, 123–124
    female, 123, 123f
    functions, 122
    male, 122–123, 123f
respiratory system,
    105–107, 106f
    disorders, 107
    functions, 105
    gas exchange, 106f
    macroscopic anatomy,
        106
    microscopic anatomy, 105
restriction point, 131, 140
Rhizopus stolonifer, 21
ribosomal RNA (rRNA),
    171, 171f
ribosomes, 23, 30
RNA (ribonucleic acid), 42, 43,
    44f, 50, 170–171, 171f
rough endoplasmic reticulum
    (RER), 23–24

sagittal plane, 88, 126
saprotrophs, 218, 223

scanning electron
microscope (SEM), 11
scavengers, 218–219, 223
science
subdivisions of, 2–3
as way of thinking, 1
scientific method, 5, 13
scientific thinking, 6
importance of, 2
process of evaluating
information, 1–2
process of investigation, 1
scleroderma, 122
S-curve of population
growth, 246, 246f
secondary succession, 234,
234f, 242
selective breeding, 143, 151
selective permeability, 55,
55f, 56f, 66
semiconservative, 172, 180
sensory neurons, 114
serous membranes, 89, 126
sex-linked disorders, 191
sexually transmitted diseases
(STDs), 123
sexual reproduction, 3,
129–130, 140
shingles, 100
shiver, 227
short bones, 93
sickle cell disease (SCD),
189, 198
simple Mendelian genetics/
inheritance, 155–157
sister chromatids, 131, 140
skeletal muscle, 95–96,
96–97, 96f, 97f
skeletal system, 91–95
diseases/disorders, 94–95
functions, 91
microscopic anatomy of
bone, 92, 92f
regions of, 93, 93f
skin/integumentary system,
98–100
of baby, 138
diseases and disorders,
100
functions, 98

macroscopic anatomy,
100
microscopic anatomy,
99, 99f
small intestine
wall, cross section of,
108f
smooth endoplasmic reticulum (SER), 23, 24
smooth muscle, 95, 96f
solute, 58, 66
solutions, 58, 66
osmotic, 59, 59f
solvent, 58, 66
species
defined, 226, 242
endangered, 281, 285
extinct, 281–282
interactions, 219–220
invasive, 282, 285
keystone, 219, 223
native, 282, 285
pioneer, 233
relationships between,
213–214
spirillum, 19
stoats, 248
strata, 265, 273
stratification, 265, 273
stress, 249
stroma, 74, 84
substrate, 45, 50
succession
communities and,
232–234
primary, 233–234, 234f,
241
secondary, 234, 234f, 242
superior section, 88, 126
sustainability, 283, 286
sweat glands, 100
symbiotic relationships, 221,
223
synapsis, 135, 140
systemic lupus erythematosus,
122

table, 7, 7f, 13, 247
table salt, 35
taiga, 230

taxonomy, 18, 19
Tay-Sachs, 190, 199
telomeres, 136–138, 137f,
140
telophase, 132, 140
temperate forest biome,
229–230
tendons, 96
terrestrial biomes, 229
testes, 122
tetanus, 97
theory, 2, 13
theory of natural selection,
264–265, 264f, 268,
273
thylakoid membranes, 73,
84
Tinea pedis, 21
tissues, 225
toxicogenomics, 202, 210
traits, 144–145, 151
dominant, 156
recessive, 156
transcription, 174, 174f,
175f, 180
transfer RNA (tRNA), 171,
171f, 175, 175f
transforming principle, 167
transgenic organism, 206,
210
translation, 174, 176f, 180
transmission electron
microscope (TEM),
10–11
transport proteins, 60, 66
transverse plane, 88, 126
trisomy 21, 194, 199
Trypanosoma species, 20
Turner syndrome, 194, 199
type 1 diabetes, 118
type 2 diabetes, 193, 199

ulcerative colitis, 110
ultrasound, 195, 199
urban sprawl, 279, 286
urethra, 122
urinary system, 110–113,
112f
diseases, 112–113
functions, 110

macroscopic anatomy,
    111
microscopic anatomy,
    111, 111f

vaccines, messenger RNA,
    204–205, 210
vacuoles, 25, 30
variables
    control, 7–8, 12
    defined, 7, 13
    dependent, 8, 12
    independent, 8, 12
variation, 155, 165
vectors, 221, 223
veins, 103

Venn diagrams, 216
venous system, 104f
ventral cavity, 89, 126
venus fly traps, 248
vertebrates, 22
vestigial structures, 266,
    266f, 273
Vibrio cholera, 20
viruses, 4
visceral membranes, 89, 90f,
    126

water, 37–38
    as pH buffer, 38
water-based ecosystem,
    279–280

water cycle, 238, 238f
Watson, James, 168
wetlands, 229
white blood cells, 102
Wilkins, Maurice, 168
Wilson's disease, 63

x-axis, 8, 247
X chromosome, 191

y-axis, 8, 247
Y chromosome, 191
yellow bone marrow, 93

zygote, 129, 140

Notes

Notes

Notes

Notes

Notes

Notes

Notes

Notes

Notes